Underwater Acoustics

Edited by

R. W. B. STEPHENS

Reader in Acoustics, Imperial College, London

WILEY - INTERSCIENCE

a division of John Wiley & Sons Ltd

London New York Sydney Toronto

Library of Congress Catalog card number 70-122350

ISBN 0 471 82204 3

Printed in Great Britain by
Adlard & Son Ltd, Bartholomew Press, Dorking

Preface

There has been a considerable growth of scientific interest in underwater acoustics during recent years and correspondingly the applications of acoustic technology have become increasingly significant. The result of this rapid technical development and of underwater exploration has been a notable augmentation of the general literature on underwater acoustics. The inspiration for producing yet another book on the subject was provided by the presence together at Imperial College, in 1967, of three leading underwater acousticians from overseas spending a sabbatical year at the College. Each of these contributed a talk on his underwater topic of interest in a series of lectures given at Imperial College to which also British workers gave specialized talks.

As the editor of the book containing these lectures, I would like to record my sincere thanks to the various authors for the excellence of their contributions and for their invaluable co-operation. Also to the publishers I express my gratitude for their helpfulness in the various stages of production of this volume.

R. W. B. STEPHENS

Contributing Authors

W. F. Hunter *Department of Physics, Imperial College of Science and Technology, Prince Consort Road, London. (Now at R.A.N. Research Laboratory, Sydney, Australia).*

R. W. G. Haslett *Kelvin Hughes Limited, Hainault, Essex, England.*

D. M. J. P. Manley *Sheldons Farmhouse, Hook, Hampshire, England.*

H. Medwin *Department of Physics, Naval Postgraduate School, California, U.S.A.*

B. Ray *Department of Physics, Imperial College of Science and Technology, Prince Consort Road, London, England.*

R. W. B. Stephens *Department of Physics, Imperial College of Science and Technology, Prince Consort Road, London, England.*

V. G. Welsby *Electrical Engineering Department, University of Birmingham, Birmingham, England.*

A. O. Williams *Physics Department, Brown University, Providence, Rhode Island, U.S.A.*

Contents

1

The Sea as an Acoustic Medium

R. W. B. Stephens
Imperial College of Science and Technology, London

1.1 Introduction

In this opening chapter it is hoped to provide a general background of information for the succeeding specialist contributions and to deal essentially with those features of the ocean which are of significance when it is considered as an acoustic medium.

The sea as a liquid medium differs from the macroscopically quiescent water, as normally met in laboratory experiments. The so-called 'unrest' of the live ocean has attracted the interest of many generations of scientists and common agreement has pointed to the need of making continuous and accurate measurements of the various physical properties. In pursuance of this fact, a recording tidal gauge was installed first in the Thames at Sheerness nearly one hundred and forty years ago, and it is interesting to note that the mechanical analysis of such records was initially made possible by the development of Lord Kelvin's harmonic analyser. A dramatic leap forward in tackling this problem of 'variability' has come about with the advent of electronic techniques in measurements and in data processing; the first international symposium on the subject taking place as recently as 1966.

Improved instrumentation has revealed that the spatial aspects of the 'variability', the stratification of the ocean, has a more complicated and fine structure than previously anticipated. For example, temperature and salinity have been found in certain regions to change quite sharply over a few metres vertically. Spectral analysis of these records reveal certain general dynamical features such as the existence of primary motions with tidal, inertial and seasonal related periods. These are 'immersed' in a continuum of periodicities which are related to changes of temperature, wind stress and barometric pressure, and there are also non-periodic movements of uncertain origin. If the relationships between these variables and sound propagation can be understood then it should be possible to forecast the principal characteristics of weather over the ocean from combined oceanographic and atmospheric measurements.

The details of the layered structure of the ocean change both diurnally and seasonally, and the uppermost layer (of the order of two hundred metres) macroscopically has an isothermal character but microscopically is broken up by turbulence into innumerable random patches, the smaller of these being controlled by the thermal diffusivity of the water. Below the isothermal layer there is the thermocline, a region where there is a large vertical temperature gradient. The temperature decreases to a minimum of 4°C at about 10,000 m when it starts to increase uniformly until the sea bed is reached. The layers of differing temperature and density may be sufficiently 'dense' to deviate appreciably the direction of incident sound waves, and moreover the layers are not necessarily parallel to the surface but may sometimes follow the sea-bed contour, and hence the importance of a knowledge of the disposition of these layers in underwater acoustic transmission.

J. D. Woods[1] (Meteorological Office) and co-divers have used dye tracers with considerable success to reveal the structures of the thermocline. To show up the existence of internal waves, several thermocline sheets were dyed by tying a number of packets of fluorescein to a horizontal line. Fig. 1.1 shows a 'jet' of fast-moving water in a thermocline sheet during the process of being sheared by a steep internal wave. The inset shows the dye drawn out by the 3 cm thick jet which lies just inside the lower edge of the 10 cm thick sheet.

Since the microstructure of the ocean changes with time, then a statistical description becomes essential and according to Skudrzyk the Kozmogorov† distribution law of turbulence may be used to describe the fluctuations of

† *Doklady Akad Nank SSSR*, **30**, 299 (1941).

temperature. This law is expressed by $E(k) = Ak^{-5/3}$, where A is a constant and k is the space wave number corresponding to the energy $E(k)$, and leads to a cube root distance variation. Pharo found experimentally that the r.m.s. temperature fluctuations do decrease with the cube root of the spacing between the thermistors measuring the temperature changes[2–5].

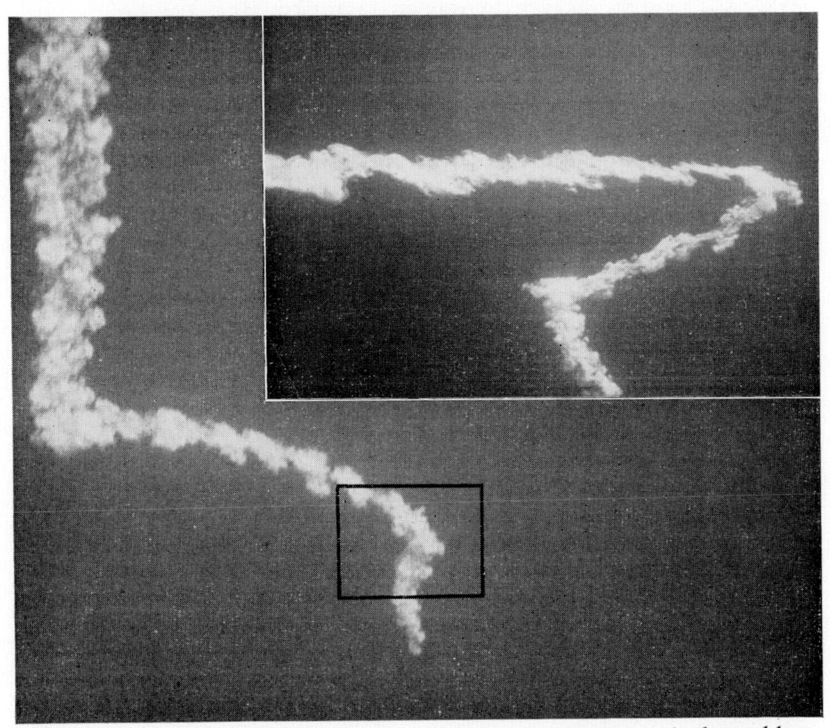

Fig. 1.1 A 'jet' of fast-moving water in a thermocline sheet that is sheared by a steep internal wave. The inset shows the dye drawn out by the 3 cm thick jet which lies just inside the lower edge of the 10 cm thick sheet. This illustration is Crown Copyright and is reproduced by permission of the Controller, H.M.S.O.

The sea contains traces of over forty elements and a cubic mile of the ocean has been estimated on average to contain 120 million tons of common salt, 18 tons of magnesium chloride, 8 million tons of magnesium sulphate, etc. In addition there are small amounts of precious metals such as gold (25 tons) and silver (45 tons). The salinity, or salt content, of sea water serves to identify a particular water mass and is the starting point for the calculation of ocean currents, sound velocity, etc. Electrical conductivity measurements provide a rapid and accurate means of determining changes of salinity.

Variations of temperature and salinity will affect, amongst other properties, the speed of sound propagation in the ocean and this is an important physical property of the sea.

Formulae[6-8] have been proposed to express this speed at zero depth (i.e. atmospheric pressure). Wilson proposed the formula

$$c = 1449{\cdot}2 + 4{\cdot}623\,T - 0{\cdot}0546\,T^2 + 1{\cdot}391\,(S-35)$$
$$+\text{higher order terms in } T^3,\ (S-35)^2,\ (S-35)\,T$$

where c is expressed in m s^{-1}, T is temperature in °C and S is salinity in parts (by weight) per thousand.

The speed of sound in the sea increases with temperature, salinity and depth as indicated in the table below.

Table 1.1 *Variation of sound speed with temperature, salinity and depth (pressure)*

Variable	Coefficient	Coefficient
Temperature (ambient about 20°C)	$\dfrac{\Delta c/c}{\Delta T} = +0{\cdot}0018$ deg^{-1}C	$\dfrac{\Delta c}{\Delta t} = +2{\cdot}7$ m s^{-1} deg^{-1}C
Salinity	$\dfrac{\Delta c/c}{\Delta S} = +0{\cdot}0008$	$\dfrac{\Delta c}{\Delta S} = +1{\cdot}2$ m s^{-1}
Depth	$\dfrac{\Delta c/c}{\Delta H} = +11{\cdot}3 \times 10^{-6}$ m^{-1}	$\dfrac{\Delta c}{\Delta H} = +0{\cdot}017$ s^{-1}

c=speed of sound in m s^{-1}, T=temperature in °C,
S=salinity in parts (by weight) per thousand,
H=depth in metres.

The physical state of sea water requires for its definition the independent measurement of three physical parameters which are usually temperature, salinity (or electrical conductivity) and pressure. Alternatively, due to its present day increased accuracy (± 10 cm s^{-1}) of *in situ* measurement, sound velocity may now replace pressure which is the less accurately determinable parameter. In circumstances when it is possible to allow the measuring system to attain the ambient temperature, a second sound velocity measurement of a sample of surface water (of known properties) contained in a flexible container may be combined with the conductivity and other sound velocity measurements.

1.2 The shape of the sea

The bed or floor of the ocean is conveniently divided into three parts, the shelf, the slope and the abyss. The shelf is the shallow platform around the edge of the land which in its widest areas, e.g. the North Sea, the Ground Banks of Newfoundland, etc., may extend for a hundred miles or so, while it may only be a few miles wide as off the steep coasts of Chile or northern Spain. The floor of the shelf has a small gradient down to its edge which is on average 60 to 80 fathoms below the sea surface, but here the gradient increases rapidly to approximately 1 in 20 or 30 down to depths of the order of 1500 to 2000 fathoms. The development of a more gentle slope marks the region of the abyss which underlies nearly eighty per cent of the Earth's sea surface. It is only as a result of sonic soundings that the real nature of the sea bed has been determined, and has revealed in particular that the slope contains many valleys and chasms and that large mountains and trenches are to be found in the abyss. These valleys and chasms are known as canyons and could have typical dimensions of 10 to 15 miles in width and a mile deep which are very similar dimensions to those of the Grand Canyon of Arizona[9].

It is evident that the variety in the topography of the sea bed is very similar to that of the land and detailed knowledge of it is essential both from the point of view of underwater communication and also of actual voyaging below surface as has been suggested is a possibility for the future.

1.3 The wave system of the sea

The wave system of the sea is very complicated and is created by various agencies such as the wind, earth movements, the sun and the moon. The waves are chiefly located near or at the surface but so-called internal, or submarine, waves do also exist. In the simple conception of the surface wave the 'particle' motion (i.e. that of a small elementary volume of the medium) will be a circular orbit in a plane perpendicular to the surface and giving no resultant movement forward with the waves. The particle trajectory becomes an increasingly flattened oval shape when the surface wave is very long compared to its vertical displacement, so that for very long waves as in ocean tides, having wavelengths of some hundreds of miles but only displacements vertically of a few feet, the horizontal traverse may reach several miles so resulting in the motion to-and-fro of the tidal currents.

Wind generated waves are more trochoidal than sinusoidal in shape, and with continued action by the wind they grow higher at a faster rate than their increase in length. When the ratio of height to length reaches about 1 to 7, instability occurs and the crest topples forward and becomes what is known as a 'white-cap.' In this way the height of the waves becomes reduced and allows the length to increase again until more white-caps are produced. Since the speed of the gravity waves depends on their length, the velocity will increase until it approaches that of the wind. It is evident that such a wave system will depend for its development on the sea distance

—the so-called 'fetch'—over which the wind can operate on the sea. The longest 'fetch' is in the southern ocean where the 'Roaring Forties' operate and here waves of 60 ft high are obtained, having crest to crest distances of several hundred yards. Cornish observed at Bournemouth at the beginning of the century waves having a period of 22·5 seconds, which corresponds to a speed of approximately 70 knots and a length of over 800 yd.

A great deal of information about the component waves making up the complicated motion of the sea has been obtained by means of pressure recorders on the sea bed and the consequent analysis of the recordings. The longest waves are first recorded because of their greater speed but their amplitude is smaller, which is explainable in terms of their loss of energy through setting a comparatively tranquil ocean into activity. The rate of advance of the 'swell' of such a wave system, i.e., the group velocity, is only half the speed of the individual waves. Hence the speed of the swell is of the order of 30 knots which is nearly double the speed at which storms move across the Atlantic so that observations of the wave system can give advanced information of the approach of distant storms.

The huge tidal waves produced by underwater earthquakes are known as 'tsunami' and are long surface gravity waves whose components, which are long compared with the ocean depth, travel with much greater speeds ($\sim$400 to 500 knots) than the shorter wave components. Although the maximum initial tidal wave crest may be as much as 80 ft in height, at a distance of a hundred miles or so away from the epicentre it will have receded to some 2 or 3 ft. However, on reaching a coastline the various phenomena such as reflection, resonance, etc., can lead to a heaping up of the wave to perhaps as much as 40 ft above the highwater mark. It is evident that useful advanced information can be achieved by the use of seismic-type detectors on the ocean bottom at strategic points to give a warning of disturbances of the sea bed[10–13].

1.4 Sound propagation undersea

Sound may be propagated within the ocean for much greater distances than other forms of energy, the detectable ranges being as great as several thousand kilometres. An experiment was performed in 1960 in which a series of explosive charges were detonated near Australia in the Pacific Ocean. The sound travelled along one of the 'sound channels' which exist under the ocean and the waves were 'guided' towards the Cape of Good Hope and then northwards in the Atlantic Ocean, to be finally received at a listening station in the Bermudas, after a journey of 13,000 miles in about three and three-quarter hours.

The sea is actually a better medium than air for range of sound transmission, and moreover it has a specific acoustic impedance (i.e., product of density and sound velocity) several magnitudes greater than that of air (and other gases). This impedance of the sea is only a little less than that of the solid materials used for piezoelectric transducers and so the latter may be designed to possess an internal mechanical impedance which approaches the radiation impedance of the loading medium. In this way a transducer used in water can have as high a conversion efficiency as 50 per cent, over an octave band.

Sound transmission under water, utilizing submarine bells and listening hydrophones, was already being practised at the beginning of the century, but no significant advancement took place until World War I, when the British system of ASDIC was developed (Anti-Submarine Detection System). This echo-ranging system had been conceived some two years before the start of World War I, following the sinking of the *Titanic*, by L. F. Richardson but he did not follow up his own proposals. Further improvement took place in World War II, when the word SONAR (SOund NAvigation and Ranging) replaced ASDIC to describe the method or equipment for determining by underwater sound the presence, location or nature of objects in the sea. It is the high velocity of sound in water, about 4·5 times that in air, which facilitates the measurement of the echo times in sonar systems, and furthermore, being about 100 times that of the maximum speed of ships, it means that the Doppler frequency shifts are of an ideal magnitude.

As yet there is no accurate theory by which sound velocity in a liquid may be calculated but an estimate can be made by considering the liquid molecules as a loosely packed assembly of rigid billiard balls. The mechanical momentum of the sound wave is considered to be transferred from one molecule to another with the kinetic gas velocity but effectively the momentum is transferred from centre to centre of the molecules. Following this argument leads to the expression

$$c_l \approx 4{\cdot}3 \left[\frac{c_v}{1-\delta}\right]$$

where c_l and c_v are the speeds in the liquid and vapour phases respectively. δ is the so-called packaging factor, being the ratio of the actual volume occupied by one liquid molecule to the volume of unit cell (i.e., the total macroscopic volume of the liquid divided by the total number of molecules).

Of particular interest to sound propagation in the ocean is the problem of the velocity of propagation in a liquid containing gas bubbles.

1.4.1 Velocity of sound in liquids containing gas bubbles

The velocity of sound in liquids containing gas bubbles is considerably changed by their presence and Karplus showed by experiment (Fig. 1.2) that the velocity varies from 1500 m s^{-1} for pure water free from gas bubbles to 20 m s^{-1} for water containing 50 per cent volume concentration of air. For a constant mass concentration the velocity is found to be directly proportional to the ambient pressure over an appreciable pressure range; consequently a slow pressure rise in such a medium becomes a shock wave very rapidly. On the other hand shock waves can be expected to be severely attenuated on passing through the medium. The heat interchange between

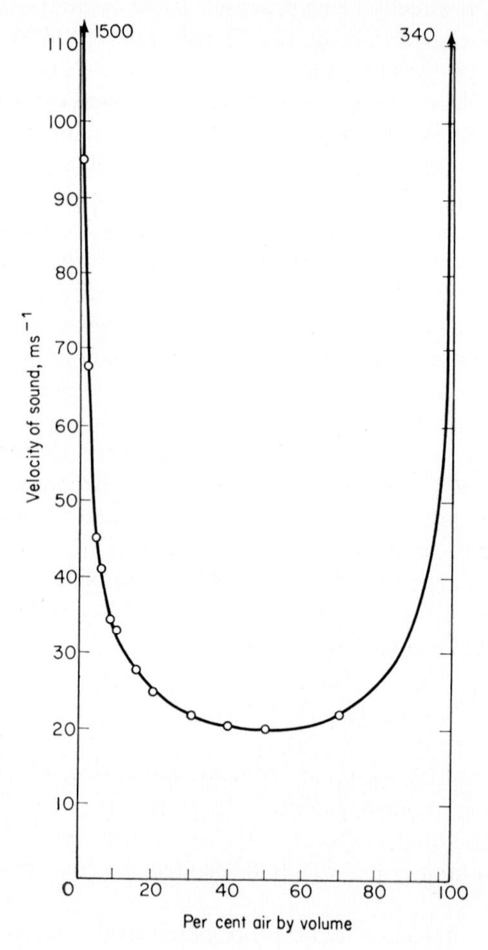

Fig. 1.2 Velocity of sound in water containing air bubbles. (After Karplus)

the gas bubbles and the water is very rapid so that there is no adiabatic temperature rise due to the compression. The formula finally developed for the velocity (c) in the medium by Karpus[13] (following a similar equation developed by A. B. Wood)[14] is

$$\frac{1}{c^2}=\frac{x^2\gamma}{c_g^2}+\frac{x(1-x)}{P}\rho_L+\frac{1}{c_L^2}$$

where the ratio (γ) of the principal specific heats for air $=1{\cdot}4$, c_g the velocity of sound in air $=340$ m s^{-1}, c_L the velocity of sound in water $=$ 1500 m s^{-1} and ρ_L the density of water $=1{\cdot}0$ g cm^{-3}. x denotes the volume concentration of gas, i.e., the ratio of volume of gas to volume of mixture. The approximations introduced only give an error $<0{\cdot}14$ per cent for $x=0{\cdot}99$ and $P=1$ atmosphere.

From the point of transducer design the acoustic impedance of the

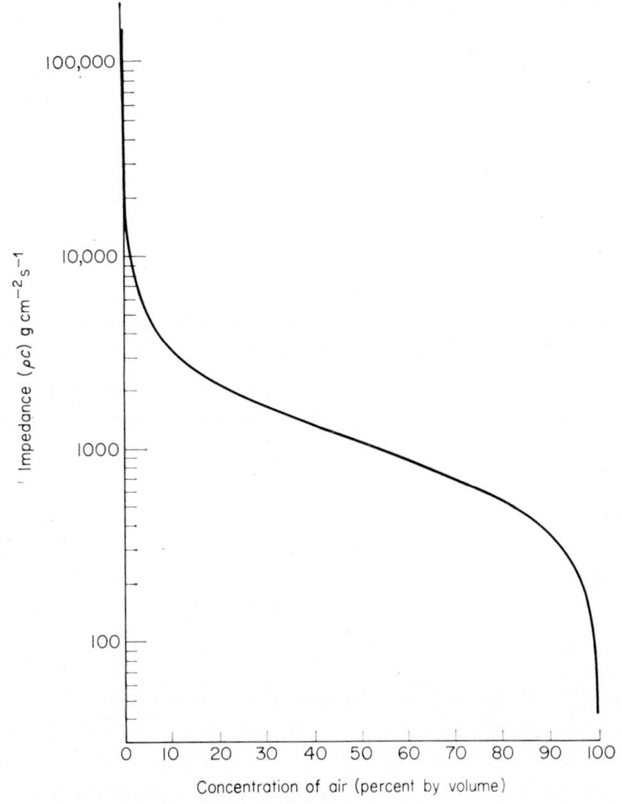

Fig. 1.3 Acoustic impedance of water–air mixture. (After Karplus)

medium is important and the graph (Fig. 1.3) shows its monotonic variation from gas-free water to air alone.

1.4.2 Attenuation

When sound is propagated within the ocean the loss of energy as transmission proceeds outwards from the source will include the geometric spreading loss, the loss of energy due to absorption in the medium and the loss or gain due to reflection or refraction. The geometric loss for spherical transmission will lead to a 6 dB loss as the distance is doubled, but under certain conditions in the sea waveguide-type propagation, as previously mentioned, can occur and results in a halving of the dB loss due to spreading.

It is appropriate here to consider in more detail the fundamental causes of the absorptive loss in a fluid medium, such as the ocean.

If the strain in a physical system is dependent not only on the stress but also on its time derivatives and if the deformation is periodic and the strain differs in phase from the stress or pressure, then mechanical relaxation is said to take place. Such phenomena are evident in the various transport processes of energy, momentum and mass transfer.

In these respective cases whenever a temperature, velocity or concentration gradient persists for a time interval which is short enough for a flow to take place, but not to attain a steady state, then relaxation will occur. An example of such a phenomenon occurs in the heat conduction between the compression and rarefaction regions of a longitudinal sound wave in an infinite medium. In this case the separation distance between the 'high' and the 'low' regions is half-a-wavelength ($\lambda/2$) of the sound and will decrease with increase of frequency, and this accounts for the increase of heat conduction loss with frequency. The presence of shear viscosity (η) is another cause of attenuation in a liquid, the relative movement of the neighbouring parts of the medium due to the passage of the sound wave being resisted by viscous forces, leading to energy loss to the wave. The total energy loss will be given by the attenuation coefficient (per unit length of path)

$$\alpha = \alpha_{\text{viscous}} + \alpha_{\text{thermal}} = \frac{2\pi^2}{\rho_0 c^3}\left[\frac{4}{3}\eta + \left(\frac{\gamma - 1}{\gamma}\right)\frac{K}{C_v}\right] f^2$$

which is obtained by additive combination of the separate attenuation coefficients, α_{viscous} and α_{thermal}. ρ_0 is the density, c the velocity of sound, γ the ratio of principal specific heats, η the coefficient of shear viscosity, K the coefficient of thermal conduction, C_v the specific heat at constant volume and f is the frequency. In the case of water $\alpha_{\text{thermal}}/\alpha_{\text{viscous}}$ is

~0·001 so that the thermal conduction loss may be neglected. The measured value at approximately 20°C of the 'attenuation' (α/f^2) in fresh water, viz. 25×10^{-17} cm^{-1} s^2, however was found to be considerably in excess of that calculated assuming only viscous dissipation, i.e., $8{\cdot}5 \times 10^{-17}$ cm^{-1} s^2.

On a macroscopic basis Liebermann[15] explained satisfactorily this discrepancy as due to the omission of the volume (bulk or dilational) viscosity, but this explanation does not provide any mechanism by which this viscosity coefficient may be calculated. Hall however, has suggested that the bulk viscosity arises from a structural rearrangement of the water molecules which can exist in two energy states, the higher one being that of closest packing. On the passage of a compressional sound wave some molecules break their structural bonds and move from the normal to the close-pack arrangement. This process will involve a time lag and hence acoustic absorption takes place.

Wilson and Leonard[16] determined the acoustic absorption in sea water contained in a glass sphere, suitably excited into a natural mode of vibration, and thereby avoided the difficulties of undersea measurements. They found that between 5 and 50 kHz the sea water absorption was 30 times that of distilled water.

This large excess of attenuation in sea water was presumed to be due to a relaxation mechanism. Leonard discovered experimentally that it resulted from the presence of magnesium sulphate, which is present however only in a concentration of about 0·17 per cent. In order to understand a little more about this energy loss the problem will be looked into more closely.

The sound absorption in aqueous solutions of some electrolytes and also in sea water is considerably in excess of that of the solvent. The frequency dependence of this absorption indicates that the overall absorption is given by the superposition of the absorption due to the solvent itself and the relaxation absorption arising from the electrolyte (Fig. 1.4).

The number of relaxation processes as revealed by the maxima of a $\mu\,(=\alpha\lambda)$ versus frequency (f) curve, has been shown from systematic investigations to depend on the type and valency of both the ion partners. μ is the attenuation per wavelength of the acoustic wave and is a more useful coefficient than α. Nearly all the 'weak' electrolytes show just one relaxation maximum but the number of maxima varies for the 'strong' electrolytes. Strong 1–1 valency electrolytes, e.g. NaBr, show no electrolyte absorption or even a small negative value. This has been explained as due to the influence of the ions on the structure of the solvent, which is apparently to partially destroy the ice-like configuration of the water molecules. By contrast strong 1–2 electrolytes show a relaxation maximum at high frequencies (~100 MHz) but for strong 2–2 electrolytes there exists a further maximum at lower frequencies. $MgSO_4$ has one at 140 kHz (Fig. 1.5), which however does not exist either in Na_2SO_4 or $MgCl_2$. These facts imply that the absorption is not caused by interaction between the ions and the solvent but that there is a specific interaction between different ion

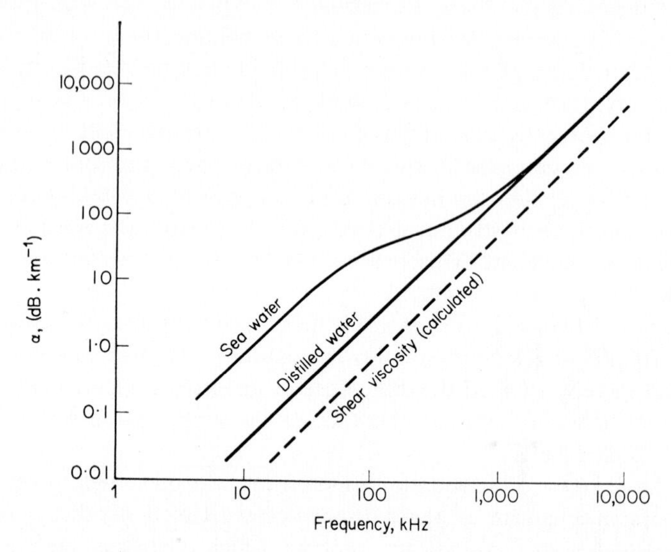

Fig. 1.4 Absorption coefficient measured in sea water and in distilled water, and calculated value assuming absorption due to shear viscosity alone

partners. The lower of the maxima for $MgSO_4$ may be ascribed to the first step of dissociation which leads from a direct contact of the ions to a complex with one water molecule inserted between the ions. This process takes place at the lower relaxation frequency.

$$Mg\overset{H}{\underset{H}{O}}\;\overset{H}{\underset{H}{O}}\;SO_4 \underset{\substack{\text{2nd step}\\\text{(Higher relaxation}\\\text{frequency)}}}{\rightleftharpoons} Mg\overset{H}{\underset{H}{O}}\;SO_4 \underset{\substack{\text{1st step}\\\text{(Lower relaxation}\\\text{frequency}}}{\rightleftharpoons} MgSO_4$$

The experimental determination of attenuation in the ocean is hampered by the very great practical difficulties of making valid measurements at sea and of the interpretation of the results. It is not an easy matter to disentangle the loss arising from energy absorption and that due to the geometrical effects of reflection and refraction, which cause the sound to be transmitted from one point to another over a multiplicity of paths. From a vast amount of data between 1 kHz and 10 MHz the following

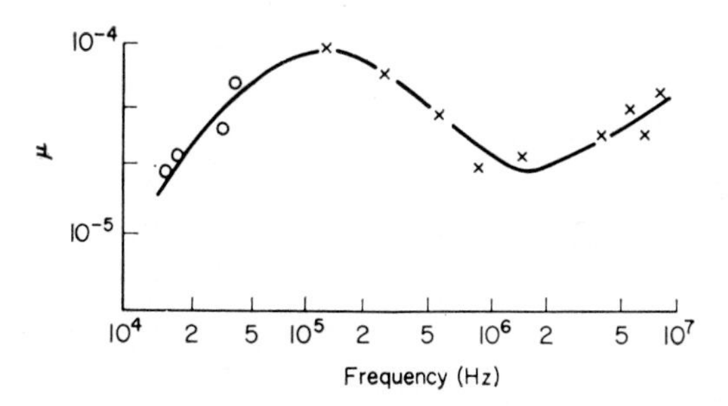

Fig. 1.5 Relaxation in magnesium sulphate solution showing two peaks. (After Tamm and Kurtze[17])

empirical equation for the attenuation coefficient α of sea water has been evolved

$$\alpha = 8{\cdot}686 \left[\frac{2{\cdot}34 \times 10^{-9}\, Sf^2}{(1 + f^2/f_T^2)\, f_T} + \frac{3{\cdot}38 \times 10^{-9} f^2}{f_T} \right] \times (1 - 6{\cdot}54 \times 10^{-4}\, P_s)\ \text{dB m}^{-1}$$

where f is the frequency in Hz, S is the total salinity in parts per thousand, P_s is the hydrostatic pressure (kg cm^{-2}), and f_T is the relaxation frequency given by $f_T = 21{\cdot}9 = 10^{(9-1520/T)}$ Hz where T is the absolute temperature.

The first term in the large bracket is associated with the 'salt relaxation' which is almost completely due to the magnesium sulphate contribution.

Hall[18] developed a quasi-crystalline model for water and his theory (Fig. 1.6) followed closely the observed change of absorption with temperature.

In the above consideration of the attenuation of acoustic signals during propagation through water no mention has been made of the phase characteristics of the transmission. For a complete knowledge of the manner in

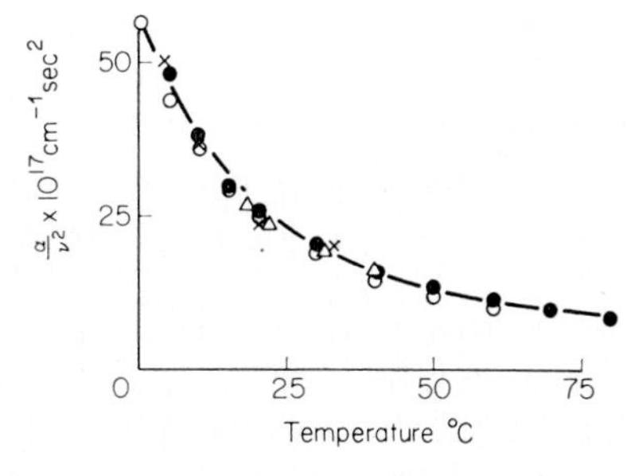

Fig. 1.6 Hall's theoretical curve for variation with temperature of sound absorption coefficient in water, together with experimental points of various observers (Pinkerton, Fox and Rock, Baumgardt, Smith and Beyer)

which any signal is transmitted a full knowledge of the transmission function is desirable. Using the electrical analogy of voltage transmission in a distributed electrical network (Bode) it is possible to synthesize a linear network whose attenuation properties simulate the response of sea water as measured by experiment. Peterson for example has developed such an equivalent network for acoustic transmissions in sea water. The impulse

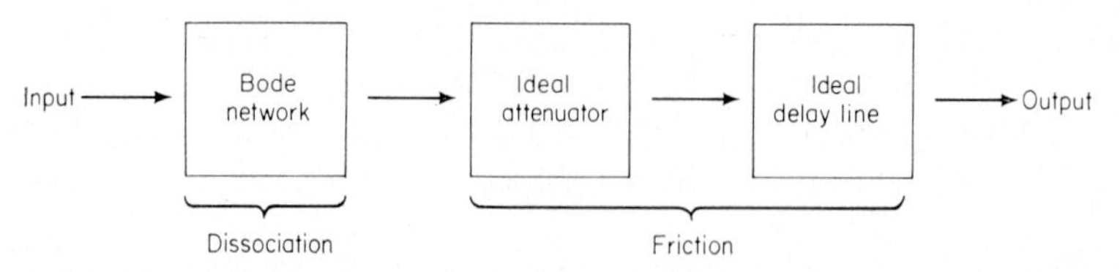

Fig. 1.7 Block diagram for sea water transmission function. (After Peterson)

response of this network having been found, it allows the calculation of the time response to any arbitrary signal in terms of the superposition integral. The total equivalent network (Fig. 1.7) comprises a minimum phase structure (the Bode network) in cascade with an ideal attenuator and an ideal delay line, these two network elements representing respectively the effects of friction and dissociation on the transmission[19–21].

1.4.3 Undersea 'noise'

Although sound waves are normally propagated through sea water with small attenuation the lower limit of detection will be influenced finally by the background noise.

Due to the wide disparity of the air and water acoustic impedances only 0·12 per cent of sound energy incident normally on to the air–sea surface penetrates the liquid medium. The ocean however is not the silent haven of popular conception as the ambient noise of the underwater environment can be quite appreciable and gives rise to problems in underwater acoustic communications. This noise may originate from marine animals (such as the snapping shrimp) which reside mainly in shallow waters, distant ships or other man-made sources, sea-surface disturbances or from water flow over rigid surfaces.

Thermal noise arising from molecular agitation of the liquid medium becomes significant in deep water and limits the upper threshold of the hydrophone to about 50 kHz. Other forms of interference, apart from the self noise of the detector system, would occur in sonar systems when transmitted acoustic energy is scattered back towards the receiver from the boundary surfaces of the ocean or internal scattering agencies.

There is also the possibility of internal disturbances affecting the thermocline and hence creating inhomogeneities in the ocean which will cause fluctuations of acoustic signals[22]. Such disturbances can occur in the form of internal waves which arise in stratified water and where the density is a function of depth. These vertically transverse waves are controlled by a gravitational restoring force, and like surface waves the small 'elementary volumes' of water move in elliptical orbits with no appreciable translational motion. This is in contrast with turbulence conditions when the energy is transported as kinetic energy of the 'elementary volumes' and at a speed which is dependent on that of the turbulent flow. The spectrum of the internal waves is very broad extending from stability oscillations of short period (one to two minutes) and amplitudes of the order of centimetres to periods of days or weeks and amplitudes attaining hundreds of metres.

Many characteristics of internal waves are explainable in terms of the two-layer model of the ocean, in which it is assumed that two stable regions of different densities are separated by an interface. Thus the phase velocity of the internal waves diminishes as the difference of density decreases, since the potential energy associated with the waves will decrease. However, dispersion and distortion are unexplained by this model and Love[23], Fjeldstad[24] and Eckhart[25] at different times proposed models with a density varying continuously with depth and also making allowance for the influence of the Earth's rotation. Such models give a normal mode solution and for each frequency there are an infinity of modes with discrete wavelengths, each being associated with a distinct amplitude–depth behaviour.

The zero mode (i.e., longest wavelength) will correspond to the surface wave and the first mode will be an internal wave with a single amplitude maximum in the ocean column[26]. The dependence of phase velocity upon wavelength will mean that the waves in the different modes will diverge even though possessing the same period. In consequence the waveforms as revealed by the isotherm will show a temporal and spatial variation. A convenient technique for such studies is by the use of thermistor chains, one chain containing up to fifteen units, which measure the temperature fluctuations at a number of fixed depths.

The actual nuisance value of the fluctuations of acoustic signals created by the inhomogeneities due to the internal waves has to be compared of course with those arising from other random inhomogeneities. For example signal fluctuations due to turbulence in any region could be of the order of 5 dB while surface image interference or air-bubble 'screens' could cause intensity changes of tens of dB. The mechanical generation of internal waves has been traced to underwater seismic disturbances while ships travelling at less than a certain critical speed can also act as sources.

The surface image interference phenomenon arises from the reflection of acoustic signals at the water–air surface of the sea and is analogous to the Lloyd fringe effect in optical reflection at a plane mirror. For large acoustic wavelengths, i.e., for frequencies < 3 kHz, the sea surface irregularities will in general be small compared with the wavelength and the surface can be regarded as a plane mirror. Moreover it is a pressure-release surface, i.e., the atmospheric air cannot sustain the sound pressure in the water, and hence an incident positive pulse will be reflected as a negative pulse. Such a set of conditions will give rise to spatial variations of intensity in the regions of overlapping direct and reflected signals as may be detected by a hydrophone. Under suitable conditions the fringe system has been observed up to 10 kHz.

1.5 Reverberation

As in room acoustics, reverberation poses similar problems for audio communication underwater. In the sea however, it is more complex since not only the boundaries return energy towards the source, but also the various inhomogeneities of the medium, such as air bubbles, marine life, suspended solids, temperature and density fluctuations. The magnitude and the location of these irregularities are moreover continually changing. However by assuming equal amplitude but random phase for the individual scatterers the amplitude fluctuations of the received energy at the source can be expected to follow a Rayleigh distribution. Such a distribution would also be expected to hold for the sea-surface reflected signals at large grazing angles, where all phases are equally probable[27, 28].

Volume scattering tends to decrease with depth at greater depths and the volume of the medium ensonified is proportional to pulse-length and beam-width, so that in the limit a very short pulse should be able to resolve the individual scatterers which are giving rise to reverberation. A significant feature of reverberation is its variability, the resulting echo received at the source from a number N of scatterers being between zero and N^2 times the intensity of an echo from a single scatterer. The random fluctuations of the received signal will depend on the relative positions of the scatterers, on the motions of the scatterers and the ship, and on the transmission loss. All these various fluctuations contribute to spectrum broadening at the receiver. The volume reverberation under the simplest conditions will decrease with the square of the range, but will be subject to attenuation by absorption in the medium, which is a function of frequency. The scattering increase with frequency is approximately 3 dB/octave. It is at short range where surface back-scattering chiefly predominates for there is a rapid decrease of surface reverberation with increasing distance, which may vary as the fourth power of the range. It is thought that scattering at low grazing angles is due to a layer of scatterers just below the sea surface so that a strongly focused beam striking the surface might give rise to three reverberation peaks, viz. one due to the surface and the others due to the layer on the outgoing and returning paths respectively.

In oceanography the term 'scattering layer' refers to dense layers of minute marine animals or organisms which cause the deflection and the dispersion of impinging sound waves. These effects lead to the occurrence of false echoes, as was found during World War II when submarines were being hunted by surface vessels. These layers vary considerably in position, moving upwards at night and downwards during daylight (or moonlight) conditions.

Reverberation from the sea bed (bottom reverberation) will in general be predominating in shallow waters. The intensity of the back-scattering will depend on the particle size distribution of a sandy and silty bed and upon its general contour. The slower rate of increase of scattering at large grazing angles is considered to arise from an increased energy loss in the ocean bottom.

In general it is thought that the reverberation measurements could reveal a great deal about the acoustic structure of the ocean but at present there is need for further information as to how the frequency spectrum varies with the characteristics of the medium and of the measuring system. By comparison of the received reverberation spectrum with that of a transmitted pulse having a broad energy spectrum, it should be possible to deduce the frequency dependence of the acoustic cross-section for individual scatterers or groups of scatterers.

1.6 Cavitation

An important effect which can arise from the propagation of sound through a liquid is the generation of small bubbles and these are characteristic of what is termed 'acoustic cavitation'. This phenomenon is associated with large acoustic pressure amplitudes and occurs when the negative pressure maximum is of the order of the ambient pressure of the liquid. Most experiments in acoustic cavitation have been carried out by means of moderate sound fields where the pressure amplitudes never approach the limiting value at which a homogeneous liquid ruptures, i.e., the tensile strength of the liquid, which theoretically for water has been variously estimated as being 500 to 10,000 atmospheres. Hence it is necessary to assume that in such experiments inhomogeneities must have been present in the liquids. Such inhomogeneities are called nuclei or micro-bubbles. Various models have been proposed for explaining some of the complex phenomena associated with acoustic cavitation. The free nucleus model often provides an adequate description because many of the sites at which cavitation starts are in effect spherical bubbles. These must be limited in size otherwise they float and move out of the sound field and rise rapidly to the surface of the liquid. In freshly drawn tap water that has been allowed to stand for a few seconds a typical nucleus would be a bubble having a radius less than 5×10^{-3} cm. In water that has been standing for several hours an average nucleus would be a bubble of radius 5×10^{-4} cm. Several mechanisms accounting for the stabilization of the bubbles against dissolution have been suggested.

The minimum acoustic pressure amplitude P_C which must be applied to a liquid to induce cavitation is known as the cavitation threshold, and this decreases with an increasing dissolved air content of the liquid and also immediately following previous cavitation. Nucleation of a liquid medium may be also brought about by the incidence of high energy particles and Sette in particular has shown that there is a reduction of cavitation threshold in liquids following neutron irradiation. Acoustic cavitation is an effective means for concentrating energy, at least in moderate sound fields, because it transforms the relatively low-energy density of the sound field into the high-energy density characteristic of a collapsing bubble. Because of this fact that it concentrates energy into very small volumes acoustic cavitation is able to give rise to such dramatic effects as the excitation of luminosity in liquids, the erosion of solids and the initiation of chemical reactions. The Noltingk–Neppiras† theory of sono-luminescence explains the phenomenon as arising from black-body incandescence of the gas in the cavitating bubbles due to adiabatic compression during their collapse. This transformation of the low-energy into the high-energy density only arises when the motion is non-linear, which is a further characteristic of acoustic cavitation.

Blake distinguished between gaseous and vaporous cavitation. The former he observed as a more or less continuous stream of very small bubbles at the focus of a sound field, and the critical pressure for the formation of these streams was only observed in water almost saturated with gas. At a higher critical pressure he found that there was a very violent formation and collapse of very short-lived bubbles which he characterized as vaporous cavitation and could be observed in both degassed and partially degassed water. Cavitation is accompanied by an intense sound and the spectrum comprises a number of discrete frequencies superimposed upon

† *Proc. Phys. Soc.*, **63**, 674 (1950).

a continuous background which appears to arise from the violent collapse of the cavities. In acoustic cavitation the spectral lines correspond to the exciting frequency and its harmonics, but in addition there would be lines representing the non-linear radial and surface oscillations of gas-filled bubbles.

The aspects of cavitation that are of importance in underwater acoustics arise firstly from the noise created by cavitating ships' propellers which under war conditions can be an embarrassment from the point of view of avoiding detection. Secondly, there is the limit imposed on the power transmitted into water by transducers in sonar arrays, etc. For a transducer immersed in water the optimum maximum power per unit area of transmitting face before the onset of cavitation is given by $P_A^2/2R$, where R is the specific acoustic impedance for water and P_A is the ambient water pressure. If the transducer is situated near the surface this gives the value of $\sim$0·3 W cm^{-2}. This value will be influenced as indicated above by the physical condition of the water and with depth, e.g., at 30 m below the water surface it will be of the order of 5 W cm^{-2}. The amplitude of the acoustic signal which can be transmitted without cavitation will be greater if pulse excitation is employed; as the pulse becomes shorter the larger will be the increase[29, 30].

An aspect of bubble dynamics which could have a bearing on a number of interesting ocean phenomena such as cavitation and the quelling of surface wave disturbances is that of the rheology of liquid–gas interfaces. The problem concerns the effects of surface elasticity and surface viscosity on the damping of such oscillations, which has received little attention until recently. The pouring of oil on troubled waters has a biblical context and has long been known to mariners, but slightly misconceived by them. In fact an oil film on a water surface does rapidly damp out surface capillary ripples, but has a negligible effect on sea waves of greater wavelength than about thirty or forty centimetres. Benjamin[31], Oldroyd[32] and others have shown that in many physical problems the increased damping of air cavities arises through viscous dissipation in an adjoining boundary layer within the liquid rather than through the quasi-viscous dissipation in the film itself. It is the calming of the small scale surface disturbances which gives a deceptive impression of the whole wave spectrum being subdued. Strasberg and Benjamin[33] observed the response of small air bubbles to sound waves in water and noted that powerful shape oscillations were generated as a result of the instability of the radial motion excited by a symmetric sound field. These surface disturbances could be of considerable significance in the instability of a cavitation bubble during its catastrophic collapse. Another peculiar property of small gas bubbles is associated with their rate of rise through a liquid medium, the drag on a very small bubble appearing to be of the order of that on a rigid sphere. One suggestion is that a film of absorbed material endows the bubble surface with a resistance to tangential displacements and this effect on the overall motion increases rapidly with decreasing size.

A further phenomenon associated with the cavitation collapse results from the supersonic velocities attained by the cavity walls which will give rise to shock waves in the surrounding fluid. Such large amplitude waves have greater penetrating power, and sound sources of lower acoustic frequencies but producing large amplitudes are now being employed to more deeply penetrate the sea bed. Their use is particularly important in the case of some wet clays whose acoustic impedance is very close to that of water, so that little acoustic energy is reflected from such sediments. These larger intensity waves involve the consideration of non-linear acoustics in underwater transmission but this topic is dealt with in a later chapter.

References

1 Woods, J. D., and G. G. Fosberry, *The Structure of the Thermocline*, Underwater Association (of Malta) Report, 1966–67.
2 Pharo, L. C., and C. H. Fitzgerald, *System for Measuring Thermal-Gradients and the Like*, U.S. Patent 2,960,866, Nov. 22, 1960.
3 Sagar, F. H., 'Acoustic Intensity Fluctuations and Temperature Microstructure in the Sea', *J. Acoust. Soc. Am.*, **32**, 112 (1961).
4 Skudrzyk, E. J., 'Thermal Microstructure in the Sea and its Contribution to Sound Level Fluctuations', *Underwater Acoustics* (V. M. Albers, ed.), Plenum Press, New York (1961).
5 Richardson, W. S., and C. J. Hubbard, 'The Contouring Temperature Recorder', *Deep Sea Res.*, **6**, 239–244, April 1960.
6 Wilson, W. D., 'Speed of Sound in Sea Water as a Function of Temperature, Pressure and Salinity', *J. Acoust. Soc. Am.*, **32**, 641 (1960); *J. Acoust. Soc. Am.* **32**, 1357 (1960); *J. Acoust. Soc. Am.*, **34**, 1113 (1961).
7 Hays, E. E., 'Comparison of Directly Measured Sound Velocities with Values Calculated from Hydrographic Data', *J. Acoust. Soc. Am.*, **33**, 85 (1961).
8 Mackenzie, K. V., 'Formulas for the Computation of Sound Speed in Sea Water', *J. Acoust. Soc. Am.*, **32**, 100 (1960).
9 Colman, J. S., *The Sea and its Mysteries*, Bell, London, 1958.
10 Defant, A., *Ebb and Flow*, Michigan U.P., Michigan, 1958.
11 Phillips, O. M., *The Dynamics of the Upper Ocean*, Cambridge U.P., London, 1966.
12 Vine, A. C., and Volkman, 'Nomogram for Wind and Waves at Sea', *Under Sea Technology*, May 1964.
13 Karplus, H. B., and J. M. Clinch, 'Sound Propagation in Two-phase Fluids', *J. Acoust. Soc. Am.*, **36**, 1040 (1964).
14 Wood, A. B., *A Text-book of Sound*, Bell, London, 1941.

15 Liebermann, L. N., 'Origin of Sound Absorption in Water and in Sea Water', *J. Acoust. Soc. Am.*, **20**, 868 (1948).
16 Wilson, O. B., and R. W. Leonard, 'Measurements of Sound Absorption in Aqueous Salt Solutions by a Resonator Method', *J. Acoust. Soc. Am.*, **26**, 223 (1954).
17 Tamm, K., and G. Kurtze, *Acustica*, **4**, 380, 653 (1954).
18 Hall, L., *Phys. Rev.*, **73**, 775 (1948).
19 Peterson, E. L., 'An Equivalent Network for Acoustical Transmissions in Sea-Water', 3*rd Intern. Cong. Acoust.*, Stuttgart, Sept. 1959.
20 Bode, H. W., *Network Analysis and Feedback Amplifier Design*, van Nostrand, New York, 1945.
21 Gorshkov, N. F., 'Propagation of Pulses in an Elastic Absorptive Medium', *J. Acoust. Acad. Sci. U.S.S.R.*, **3**, No. 2 (1957).
22 Liebermann, L., 'The Effect of Temperature Inhomogeneities in the Ocean on the Propagation of Sound', *J. Acoust. Soc. Am.*, **23**, 563–570 (1951).
23 Love, A. E. H., 'Wave Motion in a Heterogenous Heavy Liquid', *Proc. Lond. Math. Soc.*, **22**, 307–316 (1891).
24 Fjeldstad, J. E., 'Interne Welle', *Geofysiske Publikajoner*, **10**, 1–35 (1933).
25 Eckhart, G. H., 'Internal Waves in the Ocean', *Phys. Fluids*, **4**, 791–799 (1961).
26 Rattray, M., *Internal Tide Generation and Propagation from Continental Shelves II Normal Modes*, Symposium on Internal Waves, International Union of Geology and Geophysics, XIV General Assembly, Switzerland, Sept. 1967.
27 Skudrzyk, E. J., 'Scattering in an Inhomogeneous Medium', *J. Acoust. Soc. Am.*, **29**, 50–60 (1957).
28 Eyring, C. F., R. J. Christensen and R. W. Raitt, 'Reverberation in the Sea', *J. Acoust. Soc. Am.*, **20** (1948).
29 Webster, B. J., 'Cavitation', *Ultrasonics*, pp. 39–48, March 1963.
30 Flynn, H. G., 'Physics of Acoustic Cavitation in Liquids', *Physical Acoustics*, Vol. 1 Part B, Academic, New York.
31 Brooke-Benjamin, T., 'Surface Effects in Non-Spherical Motions of Small Cavities', in *Cavitation in Real Fluids* (R. Davies, ed.), Elsevier, Amsterdam, 1964.
32 Oldroyd, J. G., 'The Effect of Interfacial Stabilizing Films on the Elastic and Viscous Properties of Emulsions', *Proc. Roy. Soc. Ser. A*, **232**, 567 (1955).
33 Brooke-Benjamin, T., and M. Strasberg, *Excitation of Oscillations in the Shape of Pulsating Gas Bubbles* (Two papers, one theoretical and one experimental), 55th meeting of American Acoustical Society, Washington, D.C. 1958.

2

Normal-mode Methods in Propagation of Underwater Sound

A. O. Williams, Jr.
Physics Department, Brown University, Providence, Rhode Island, U.S.A.

2.1 Introduction

In this chapter we outline a mathematical description of underwater-sound propagation to great ranges, using chiefly the powerful method of *normal modes*[1]. 'Great ranges' implies horizontal distances r between source and receiver that far exceed the acoustic wavelength λ or the water depth H, whichever is the more stringent requirement (usually, the latter). Such ranges can be expected only with moderately or very low acoustic frequencies, at which the absorption coefficient of sound in sea water is small (Chapter 1).

To those acquainted with practical underwater sound and the methods of ray acoustics, much that is presented here may seem abstract and remote. Therefore we first offer some justification. Undeniably, study of underwater sound is important, for non-military as well as military reasons, but it does not follow that all parts of the study are equally useful, or that some conceivable investigations would be useful at all.

In terms of aiding fundamental science, research in underwater sound will almost certainly contribute nothing to pure physics. There may be some discoveries in chemical physics; see Chapter 1 for a past example: the role of ionic relaxation processes in acoustic absorption. Recent measurements of sound attenuation in the sea at very low frequencies[2] show other anomalies; the causes are not yet known[3], but relaxation may contribute. Undoubtedly, however, any major scientific gains will occur in oceanography and geophysics.

Yet there is another reason for pursuit of some fairly abstract study (apart from the mental satisfaction of solving *any* problem), and that is to bring order out of chaos, simplicity out of complexity and interconnection out of separateness. A half-century's pursuit of underwater acoustics has established a great deal of empirical knowledge which is still imperfectly linked up by scientific and mathematical explanation. Consequently, any progress in widening our areas of thorough understanding is likely to have eventual applications.

Finally—a dictum always presented to student beginners in science—it is far better to understand two or more ways of solving a specific problem, rather than one, even though the same answer is obtained.

These are brave words, but a look at the puzzle set us by Nature is almost enough to frighten any sensible scientist or mathematician off to some more hopeful field of research. (Engineers are a different case; they are, or should have been, indoctrinated with the view that one has to do something about an assigned problem, no matter how hard it is.)

2.2 The physical problem and its idealization

Fig. 2.1 attempts to summarize the complexities besetting a study of underwater sound propagation. In it, x, y, z and t have their usual meanings of position and time; ω is angular acoustic frequency; α is the acoustic absorption coefficient for amplitude (2α for intensity); c is the speed of sound; ρ is the material density; subscripts w and b designate *w*ater and *b*ottom material, respectively. We have omitted a few influences, such as curvature and rotation of the earth, ocean currents and the inevitable presence of other sound sources (wind, flow noise, ships, biological organisms, etc.). Moreover we have accorded only scant attention to the

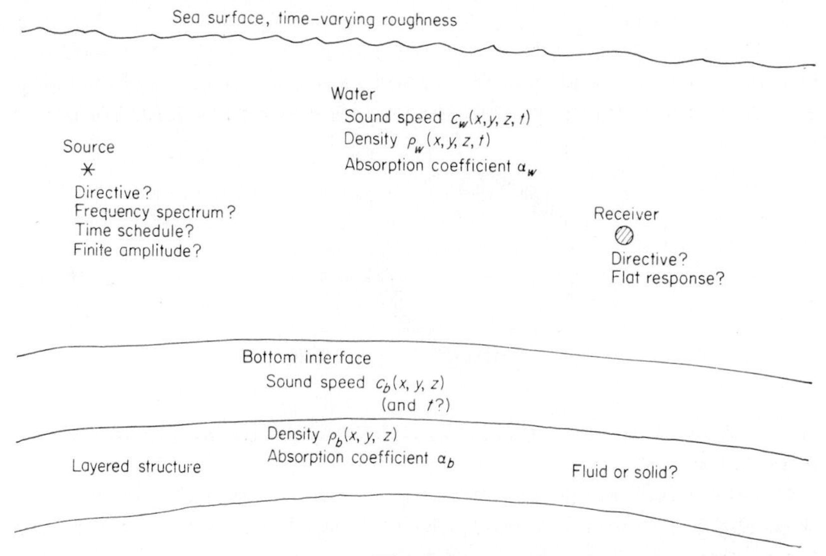

Fig. 2.1

receiver and what happens thereafter, for that is the province of other chapters.

What to do?

A professional joke that circulated among scientists a few years ago may guide us. It seems that a large dairy company had been losing money for some time. All the usual curative tactics had been tried—time-and-motion studies, more advertising, better treatment of the cows and even of the employees, shake-up in management—but to no avail. In desperation the directors voted to get a scientist as a consultant, for is not science remaking the world? Moreover, he was to be the cream of the cream among scientists, a theoretical physicist. Eventually a brilliant young theorist was found and hired at high pay. He was given a large office, telephones, a secretary, a blackboard, several assistants and unlimited access to libraries and computers—all the little amenities that make tolerable the monastic life of science. He was asked to study the company's problems and, when ready, to summon the directors to hear his analysis. In due time he did summon them, sat them down in front of his blackboard, and began, 'You have employed me to analyse your troubles and work out solutions. We scientists know that the main thing is to select first the essentials of the problem, leaving secondary matters for later corrections. I shall start (Fig. 2.2) by postulating a spherical cow . . . '.

Let this be our first task, reducing Fig. 2.1 to Fig. 2.2, as nearly as we can. Some of the simplifications to be made are demonstrably safe, some a matter of postponement for later corrections. Moreover, the nature of a specific problem may allow or dictate approximations; e.g., in certain cases[4] the ocean can be treated as if it extended infinitely downward. We use an

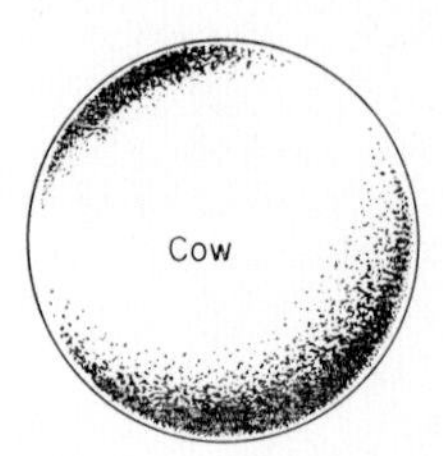

Fig. 2.2

intermediate set of approximations, commenting later on how to improve some of them.

(1) The rotation and curvature of the earth are ignored; the curvature is so slight that sound continues to be 'guided' between the upper and lower interfaces of the ocean, and the rate of rotation is too slow to have important effects.

(2) The air–water interface will be treated as a pressure-release plane surface (see Chapter 3 for some aspects of surface roughness).

(3) The depth H of the ocean is taken as constant.

(4) Variations in density of the water, important though they are in oceanography, are ignored.

(5) Absorption of sound is ignored.

(6) The speed of sound, c, in the water is a function of vertical distance z alone, and a rather simple time-independent function (see below).

(7) The bottom interface is a horizontal plane, and the bottom material is an infinitely deep, homogeneous, isotropic and non-absorbing fluid, with of course a density exceeding that of the water, and with a sound speed exceeding by a few per cent that in the water just above the bottom.

(8) The real source is replaced by a point source, radiating uniformly and continuously in all directions, at constant 'infinitesimal' amplitude and constant angular frequency ω.

(9) All other sound sources are ignored.

(10) The receiver is non-directional, and flat in its response to any frequency we use.

(11) All fluctuations in space and time are ignored.

We cannot categorically state which approximations, 1–11, are most seriously restrictive; it depends on the problem. In rather general terms, however, (1) and (4) are quite safe; (3), (5) and (8) are oversimplifications but capable of later improvement, and so perhaps are (6) and (11). Items (9) and (10) chiefly concern the signal processer (Chapter 4). Items (2) and (7) may in various cases be very grave oversimplifications, yet difficult to improve upon—particularly (7).

The assumed behaviour of $c(z)$, item (6) above, needs more exposition. Fig. 2.3 shows in solid line an average 'profile' of $c(z)$ for the North Atlantic[5]; the cross-hatched region indicates that in the top few hundred feet of the ocean $c(z)$ can behave quite differently from place to place and time to time. The physical reasons for this general shape are known.

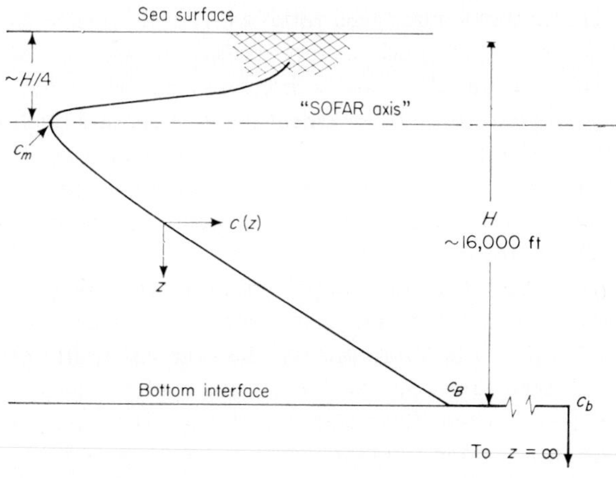

Fig. 2.3

(a) On the average, the water temperature drops as one goes down from the surface, which makes $c(z)$ decrease until finally the temperature approaches a constant; (b) but greater depth in the sea means greater hydrostatic pressure, which acts to increase $c(z)$ and eventually wins out. Other non-arctic oceans have similarly shaped profiles for $c(z)$ although, e.g., the depth for minimum sound speed may be decidedly different. Since the lower part of the profile is determined mainly by pressure alone, changing the depth H of the ocean produces an effect approximated closely by moving the bottom interface in Fig. 2.3 up or down without altering the profile; the 'jump' in sound speed going into the bottom may change, however.

We repeat the implication of Fig. 2.1, that the profile in Fig. 2.3 can change spatially, temporally and in a fluctuating sense, particularly in the upper half of the ocean. Nevertheless, we ignore these disturbing influences.

For shallow water (H not over two or three hundred feet), a good first approximation is to take $c(z)$ as constant in the water. The 'jump' in c at the bottom will now differ considerably from place to place, because the bottom material differs; a jump of 10 to 20 per cent is not unusual. Moreover, the bottom material must be treated more realistically than in our previous list, at the very least to the extent of postulating absorptive mechanisms in it[6, 7]. Almost all of our attention here will be devoted to the deep-water case, mainly because a good deal has already been published on normal-mode solutions in shallow water.

2.3 Comments on the ray method

The assumption that sound speed varies with depth z means that acoustic beams will be refracted much as are beams of light in an optically inhomogeneous medium. A *refractive index* $n(z)$ can be defined for sound, $n(z)=c_1/c(z)$, where c_1 is any convenient fixed value, and Snell's law then holds for 'rays' of sound[8]. If the tangent to a ray, at a point where the speed of sound is c_1, makes an angle θ_1 with the horizontal, then the corresponding angle θ at another point where the speed is c satisfies the relation $\cos\theta_1/\cos\theta=c_1/c=n$.

The discovery that refraction is important in sonar work, despite the fact that $n(z)$ never varies more than a few per cent, appears to have been made in the 1930s. To be more precise, the time variability of refraction was recognized then (most practical work was confined to the uppermost layer of the ocean, where the water temperature is very far from being constant with time). One consequence was development of the *bathythermograph* (BT), a simple rugged instrument to be let down a few hundred feet into the sea, from ship or submarine, recording a trace of water temperature versus depth (actually, hydrostatic pressure). When the BT was hauled up, this plot, which in effect was of $c(z)$ versus z, was recovered and studied. Then the existing sonar conditions could be estimated as good, medium or bad, and in operations some remedy might be applied, such as running the submarine at a different depth. The BT is still used in research, but for precise work much more elaborate *velocimeters* are preferred. These measure the sound speed $c(z)$ directly, as a function of depth (pressure), by timing short pulses of high-frequency sound over a path of a few inches. Precisions of better than 1 part in 5000 are attainable in c, although not in z.

2.4 Ray methods

Once the refractive index $n(z)$ is known, it is simple in theory, and sometimes in practice, to map out a predicted sound field by ray theory. A number of rays are started from the source, at known angles to the horizontal (here the directivity of the source is important, for it is a waste of time to trace rays in directions where very little sound is actually emitted). By use of Snell's law, these rays are traced forward. Whenever a ray strikes the upper or lower interface of the ocean, assumptions about its reflection must be applied. Just as is true in geometrical optics, the relative crowding or dispersal of rays in a given region is a measure of the sound intensity to be expected there.

So primitive a ray method, exactly analogous to Newton's corpuscular optics, is rarely used without elaboration. For example, the 'travel time' (of an acoustic pulse) along a calculated ray trajectory can be computed—$c(z)$ being known everywhere—and therefore the shape of acoustic wavefronts can be estimated; this in turn allows some corrections for interference

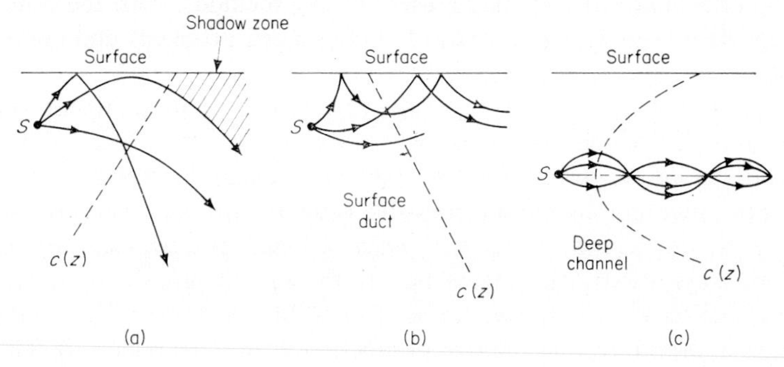

Fig. 2.4

effects. Also, ability to calculate travel time allows knowledge of whether or not sound emitted in short pulses reaches a given spot essentially simultaneously by quite different ray paths. Comparison of predicted and observed time of arrival for pulsed signals can identify the important transmission paths (direct, single reflection at the bottom, etc.).

Fig. 2.4 sketches three familiar simple cases of ray plots, with the trend of $c(z)$ shown in dashed line. In each case, S marks the source. In (a) and (b) rays incident on the surface are assumed to be specularly reflected; in (c) rays starting at sufficiently small angles with the horizontal never reach the surface (nor the bottom).

The ray method is straightforward, and it handles fairly easily a directive source (and receiver), as well as a pulsed source, but there are also disadvantages. (1) Most easy cases have been done long since, and the complicated situations of interest in present research usually require machine calculation in quantity. (2) At each incidence on surface or bottom, each ray has to be 'told' at what angle to go off, and with what percentage of total reflection. (3) Since problems are almost entirely numerical, each variation is nearly as hard as the first try: e.g., a new source depth (even a new inclination of a directive source), or a greater range (doubling the range requires approximately doubling the amount of computation).

Hence most ray problems nowadays require a computer of the order of an IBM 360, and for substantial times. Most laboratories doing such work devise their own computer programs, with their own choice of approximations, omissions and so on. Therefore, even though each finished ray plot looks clear and definitive, it is not certain that two different laboratories would always get equivalent results for the same problem. Over the last few years the Technical Committee on Underwater Acoustics of the Acoustical Society of America has been examining this latter difficulty, by enquiries addressed to major users of ray methods, with the hope of getting these users to try a series of standardized problems and compare results.

There is another shortcoming of the ray method, more pertinent to our topic. Sound, like light, is a wave phenomenon, not a ray process. The ray method derives legitimately from the wave equation, via the eikonal equation[8], in a limiting sense expressible roughly thus: *For a ray treatment to be safe, the acoustic wavelength must be much smaller than any other pertinent length.* 'Pertinent lengths' include the water depth, size of obstacles, scale of surface or bottom roughness, distance in which the refractive index changes appreciably, and the size of any predicted focal region. It follows that at low acoustic frequencies any result of a ray calculation is suspect, although by no means necessarily wrong. No exact figure for the onset of suspicion can be given; it depends on the problem being tackled. Examination of theoretical and experimental results suggests that in the deep ocean calculations at some 200 Hz and lower warrant examination by wave-acoustic theory; above some 1000 Hz, ray methods would usually be safe; in the interval, intercomparisons are desirable. Problems involving shallow-water regions, or shallow 'surface-based ducts', may require wave acoustic analysis at frequencies well above 200 Hz.

2.5 Introduction to wave theory

We now embark on the mathematical part of this discussion. The acoustic field in the ocean will be characterized by the (scalar) *velocity potential* $\Phi(x, y, z, t)$; the negative of the gradient of Φ is acoustic particle velocity, and $\rho \partial\Phi/\partial t$ is acoustic pressure. This choice of dependent variable is not essential: the *displacement potential* or the *acoustic pressure* itself can be used instead of Φ, with only minor changes in what follows. The simplifications that we have made in Fig. 2.1 allow use of cylindrical coordinates, with the vertical z axis through the point source and the horizontal range r measured from the z axis. Our assumptions imply cylindrical symmetry, so that no azimuthal angle need be specified. Choices of origin and direction for $+z$ are completely open. For definiteness, we specify positive z in the downward direction, with $z=0$ at the minimum of sound speed, $c=c_m$ in Fig. 2.3.

The acoustic-wave field is governed by the wave equation

$$\nabla^2\Phi - [c(z)]^{-2}\partial^2\Phi/\partial t^2 = 0 \tag{2.1}$$

together with proper boundary conditions at the upper and lower interfaces, at the source, and at very large distances from the source. The Laplacian operator ∇^2, or $\partial^2/\partial x^2 + \partial^2/\partial y^2 + \partial^2/\partial z^2$, is put into cylindrical coordinates

$$\nabla^2 = \partial^2/\partial r^2 + r^{-1}\partial/\partial r + \partial^2/\partial z^2 \tag{2.2}$$

Assumption (8), concerning the source, lets us write

$$\Phi(r, z, t) = \phi(r, z)\, e^{-i\omega t} \tag{2.3}$$

Eqn. 2.1 then reduces to

$$\partial^2\phi/\partial r^2 + r^{-1}(\partial\phi/\partial r) + \partial^2\phi/\partial z^2 + [\omega^2/c^2(z)]\,\phi = 0 \tag{2.4}$$

There are two broadly different avenues for solving Eqn. 2.4, as well as other intermediate ways. All come to the same result, so that the choice of method is a matter jointly of personal taste and personal background in mathematical techniques. The traditional avenue, which stems partly from much older theoretical investigations of radio transmission, is exemplified in (i) to (iv) of the five books described in the reference list. The solution of Eqn. 2.4, but with a point source already present, and with regard to boundary conditions that must be met, is written as a Fourier integral over a variable propagation constant k. This integral is then evaluated in the complex plane, which allows it to be broken up into two parts. One part is expressible as a sum of residues at poles of the integrand; these individual terms are the discrete *normal modes* of the problem. The other part is a branch-line integral, describing mainly the near field of the source; it is equivalent to a continuous set of normal modes.

The other avenue, and that to be followed here, starts by separating Eqn. 2.4 into r- and z- parts without regard to any particular source, but with full allowance for the other boundary conditions (listed below). The general normal modes are worked out, as a complete set of characteristic patterns of acoustical behaviour for the whole medium—ocean, surface and bottom. These modes are then fitted to the prescribed source, and finally the specific acoustic field is formulated as a combined sum and integral over the general normal modes, which have been individually 'weighted' in accordance with properties of the source. The mathematical procedures resemble those of Schroedinger quantum mechanics and therefore are convenient for drawing upon the vast literature of solved wave-mechanical problems. Also it can be argued (although not necessarily accepted) that this procedure is more pictorial of physical behaviour in underwater propagation. Book (v) in the reference list uses mainly an approach of this sort, with individual refinements and elaborations that open the way to tackling a variety of other associated topics, such as general ray theory and the halfway ground between ray and wave methods.

Starting as outlined in the preceding paragraph, we note that Eqn. 2.4 is separable; i.e., we can assume that

$$\phi(r, z) = v(r)\, u(z) \tag{2.5}$$

substitute vu into Eqn. 2.4, perform the indicated differentiations, divide through by vu, and rearrange the result into the form

$$u^{-1}(\mathrm{d}^2u/\mathrm{d}z^2) + \omega^2/c^2(z) = -v^{-1}(\mathrm{d}^2v/\mathrm{d}r^2) - (rv)^{-1}\,(\mathrm{d}v/\mathrm{d}r) \tag{2.6}$$

But these two functions of the independent variables z and r cannot *always* be equal to each other unless each side of Eqn. 2.6 *always* equals the same *separation constant*. This constant we write as k^2, discussing its physical meaning later.

The r-equation (right-hand side of Eqn. 2.6 set equal to k^2)

$$\mathrm{d}^2v/\mathrm{d}r^2 + r^{-1}(\mathrm{d}v/\mathrm{d}r) + k^2v(r) = 0 \tag{2.7}$$

is Bessel's equation of order zero, having solutions $J_0(kr)$, $N_0(kr)$, or any linear combination of these. The choice of solution is set by requiring outgoing waves at very large r, which demands the Hankel function of the first kind

$$v(r) = H_0^{(1)}(kr) \equiv J_0(kr) + i N_0(kr) \tag{2.8}$$

For $kr \gg 1$, this function has an asymptotic form

$$v(r) = H_0^{(1)}(kr) \sim e^{-i\pi/4}(\pi kr/2)^{-1/2}\, e^{ikr} \tag{2.9}$$

When the factor $\exp(-i\omega t)$ of Eqn. 2.3 is restored, Eqn. 2.9 represents an outgoing cylindrical wave

$$v(r) \sim (2/\pi)^{1/2}\,(kr)^{-1/2} \exp\,[i(kr - \omega t - \pi/4)] \tag{2.10}$$

The factor $r^{-1/2}$ shows cylindrical spreading; the intensity, which is proportional to v^2, falls off as $1/r$. Evidently k can be written as $2\pi/\lambda$, where λ is some acoustic wavelength, although obviously a constant rather than a function of z, as the free-field acoustic wavelength $2\pi c/\omega$ must be.

Then the condition $kr \gg 1$, required for Eqn. 2.9, merely says that horizontal ranges of interest must exceed a wavelength, or else the exact Hankel function must be retained.

The z-equation

$$\mathrm{d}^2u/\mathrm{d}z^2 + \{[\omega^2/c^2(z)] - k^2\}\, u(z) = 0 \tag{2.11}$$

closely resembles Schroedinger's equation of quantum mechanics, and the resemblance is increased if we add and subtract $\kappa^2 = \omega^2/c_m^2$ in { }

$$\mathrm{d}^2u/\mathrm{d}z^2 + \{[\kappa^2 - k^2] - [\kappa^2 - \omega^2/c^2(z)]\}\, u(z) = 0 \tag{2.12}$$

This quantity c_m can be any fixed value of the speed of sound; we choose it to be the minimum c_m in Fig. 2.3. Henceforth we use abbreviations suggested by Schroedinger's equation

$$\kappa^2 - k^2 \equiv E; \quad \kappa^2 - \omega^2/c^2(z) \equiv V(z) \tag{2.13}$$

with E corresponding to total energy, and $V(z)$ to potential energy, in Schroedinger's equation. Maintaining this resemblance will aid those acquainted with quantum mechanics and will not hurt others†. Eqn. 2.12 now becomes

$$\mathrm{d}^2u/\mathrm{d}z^2 + [E - V(z)]\, u(z) = 0 \tag{2.14}$$

2.6 The eigenvalue problem

Eqn. 2.14 is in a form that, with assigned boundary conditions on $u(z)$, has been long studied in the theory of differential equations: the *Sturm-Liouville problem*[9]‡. It is convenient to plot $V(z)$ versus z, although the frequency ω must be specified to make the vertical scale known numerically. Fig. 2.5 is the result; data for it came from Ref. 4. Setting $V(z)$ equal to infinity at, and everywhere to the left of, z_s satisfies one acoustic boundary condition. The sea surface, at z_s, was assumed to be pressure-release, so that acoustic pressure p vanishes at z_s and to the left of it. Then Φ must also vanish and, since neither the time- nor the r-part of Φ can ever be zero at finite r, it follows that $u(z)$ vanishes at $z \leqslant z_s$. The stipulation $V = \infty$ in the regions mentioned automatically makes $u(z)$ vanish there and satisfies Eqn. 2.14 by letting $\mathrm{d}^2u/\mathrm{d}z^2$ remain finite.

At the bottom interface (the bottom material having been postulated as

† There is another reason for introducing κ^2. The quantity $\omega^2/c^2(z)$ in Eqn. 2.11 varies by only a few per cent of itself over the whole ocean depth, and solution of Eqn. 2.11 or 2.12 reveals that k^2 varies even less, for most cases of interest. Subtracting κ^2 to get Eqn. 2.12 introduces smaller quantities E and $V(z)$ that vary relatively greatly, a rather more convenient behaviour than that of k^2 and ω^2/c^2.

‡ Unfortunately, our boundary conditions—to be listed shortly—are a little more complicated than those of the Sturm-Liouville problem, because the density and not merely the sound speed changes on crossing the bottom interface.

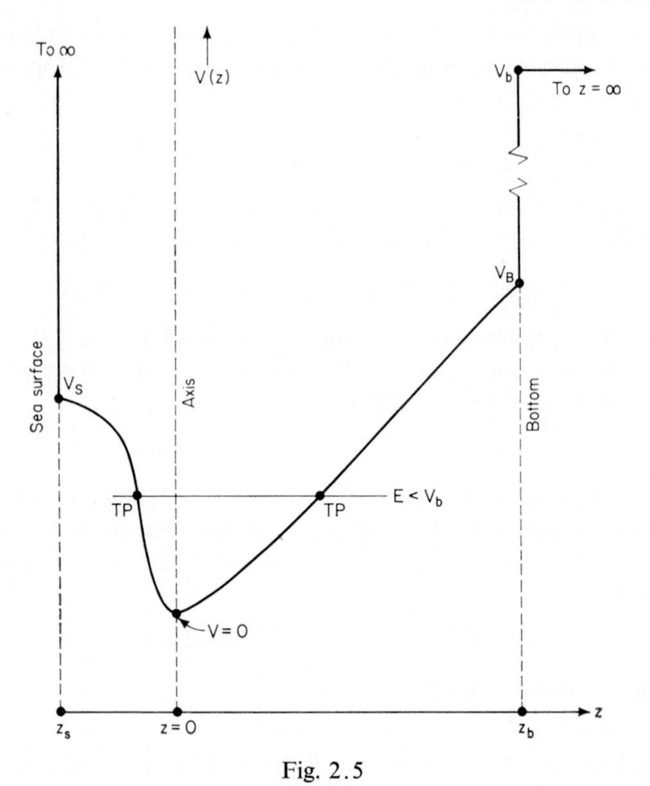

Fig. 2.5

a fluid) the acoustic boundary conditions are: (a) continuity of acoustic pressure, i.e., continuity of $\rho(z) \, . \, u(z)$; (b) continuity of z-component of particle velocity, i.e., continuity of $\mathrm{d}u/\mathrm{d}z$; and (c) in the bottom itself

$$u(z) \xrightarrow[z \to \infty]{} 0 \tag{2.15}$$

faster than $1/z$. If (c) were not true, in the assumed infinite extent of the bottom material there would be an infinite amount of acoustic energy. We have not yet tried to put in the source as part of the boundary conditions; that can wait a while.

It would take far too long to reproduce the analysis of such a *boundary-value problem*. Reference 9 will be of help, and there are brief descriptive treatments in the introductory parts of two papers by the writer[10, 11]. We shall, however, outline a naive approach, assuming that $V(z)$ is known numerically, as is the 'jump' of density ρ from ρ_w to ρ_b at the bottom interface. In many cases the ratio ρ_b/ρ_w need be known only roughly, for a reasonably good solution.

The first step is to choose a trial value of $E < V_b$, as shown in the lower part of Fig. 2.5. This line for E cuts the curve $V(z)$ at two *turning points*, TP. To the left of the left-hand TP, $E < V(z)$, and solutions of Eqn. 2.14 will be monotonic functions, roughly like $\exp(\pm \text{const. } z)$, or even more rapidly varying with z.

Starting at $z = z_s$, where $u = 0$, we integrate Eqn. 2.14 numerically, towards the right, using the selected trial value of E. For definiteness, a rising function of $(z - z_s)$ will be chosen. When the left-hand TP is reached, $u(z)$ will be continuous and so will du/dz. Beyond this TP, $E > V(z)$, and the solution changes to an oscillatory form. At the right-hand TP, u and du/dz are continuous, but now $E < V(z)$ and the solution returns to a monotonic form. Henceforth $u(z)$ must always have a slope that brings it asymptotically towards zero, to satisfy the condition that u vanishes at $+\infty$.

As the numerical integration proceeds across the bottom interface, the acoustic boundary conditions (continuity of pressure and of z-particle velocity) must be met; also a factor $\rho^{1/2}$ in $u(z)$, described below, must change. Thereafter $u(z)$ should fall as a simple decaying exponential, towards zero at $z = \infty$. At the first trial it will do no such thing, because we merely guessed at a value of E. Hence a new trial E must be chosen, Eqn. 2.14 must be integrated again and so on, until $u(z)$ behaves properly everywhere. Now we know some one combination, $[E_n, u_n(z)]$; which particular integer n is can be established by counting the nodes of u_n between the TP's.

A different E is next selected for trial, and the game goes on to find another E and u. Even with a large computer, so naive an approach would take a great deal of time; we outlined only a possible method, not the best one. Much more sophisticated tactics exist, and there are many approximate procedures that will give fairly good starting solutions, to be refined by iterative computation.

The main properties of the solutions $[u_n(z), E_n]$ will now be listed.

(1) The boundary conditions are so stringent that for $E < V_b$ (Eqn. 2.13 and Fig. 2.5) Eqn. 2.14 can be solved only for a finite number of discrete or 'quantized' values of E, E_n $(n = 0, 1, 2 \ldots N)$; each function $u_n(z)$, corresponding to one E_n, is different from all others. The E_n's are called *eigenvalues* of Eqn. 2.14 and the u_n's, *eigenfunctions*.

(2) Fixing an eigenvalue E_n determines (Eqn. 2.13) a value of k_n, which appears in $v(r)$ (Eqn. 2.9). Therefore we have a finite set of $\Phi_n(r, z, t) = \phi_n(r, z) \exp(-i\omega t)$

$$\phi_n(r, z) = \text{const}_n \times u_n(z)\,(k_n r)^{-1/2} \exp\left[i\left(k_n r - \omega t - \frac{\pi}{4}\right)\right] \tag{2.16}$$

'Const' subsumes $(2/\pi)^{1/2}$ from Eqn. 2.9 and, for the present, any other multiplying constant, because the wave equation is linear.

(3) For $V_b \leqslant E < \infty$, there is a continuous spectrum of eigenvalues E, and therefore of k's, each with an eigenfunction $u_E(z)$ differing only infinitesimally from its neighbours. These 'continuous' solutions of Eqn. 2.14 are essential, later, when a point source is introduced, but for reasons to be noted below they are usually not important otherwise.

(4) Each eigenfunction u_n or u_E must include a factor $\rho_w^{1/2}$ (ρ is density) when z specifies a point in the water, or $\rho_b^{1/2}$ when z is in the bottom. This makes $\rho^{1/2}u$ discontinuous across the bottom interface, but ϕ (and therefore u) is already discontinuous there because of boundary condition (a) given before Eqn. 2.15.

(5) When thus multiplied by $\rho^{1/2}$, the u's form an *orthonormal set*. That is, each function u can be multiplied by a constant so that for the u_n's

$$\int_{z_s}^{\infty} u_n(z)\, u_j(z)\, \mathrm{d}z = 1, \text{ when } n = j\text{—}\textit{normalization} \tag{2.17a}$$

$$= 0, \qquad n \neq j\text{—}\textit{orthogonality} \tag{2.17b}$$

For the u_E's, the corresponding expression is

$$\int_{z_s}^{\infty} u_E(z)\, u_E{}'(z)\, \mathrm{d}z = \delta(E - E') \tag{2.17c}$$

where δ is the Dirac delta-function[12]. The multiplication by $\rho^{1/2}$ is necessary to secure orthogonality[10]. Henceforth we assume that the necessary multiplying constants are already included in the u's, to satisfy Eqn. 2.17 a, b and c.

(6) All the allowed solutions $u_n(z)$ and $u_E(z)$, taken together, form a *complete set*, so that any other similarly behaved function can be expanded in terms of the u's

$$f(z) = \underset{j}{\mathbf{S}} A_j u_j(z) \tag{2.18}$$

$$A_j = \int_{z_s}^{\infty} f(z)\, u_j(z)\, \mathrm{d}z \tag{2.19}$$

Here subscript j represents either n or E as necessary; $\mathbf{S}$ indicates a sum over the 'discrete' u_n's and an integral over the 'continuous' u_E's.

(7) The lowest eigenfunction $u_0(z)$, corresponding to the lowest allowed E_0 in Fig. 2.5, has no axis-crossing or node, although like all the u_n's it vanishes at $z = z_s$ and at $z = +\infty$. Each successively higher eigenfunction

has one more node than its predecessor†—one for u_1, two for u_2, n for u_n. All these nodes occur inside the curve of $V(z)$ [i.e., between the TP's, at which E_n equals $V(z)$] where the eigenfunctions are oscillatory in character.

(8) Each combination of eigenvalue E and eigenfunction $u(z)$ represents a *normal mode* of the whole problem. A normal mode is one distinctive acoustic pattern that can exist in this (simplified) ocean and bottom. An analogy exists in the normal modes of a stretched violin string (Fig. 2.6), for which the pattern of vibration corresponds to $u(z)$, and the eigenvalue is the frequency, or wavelength, or wave propagation constant (it does not really matter which is chosen). With an origin of coordinates chosen at

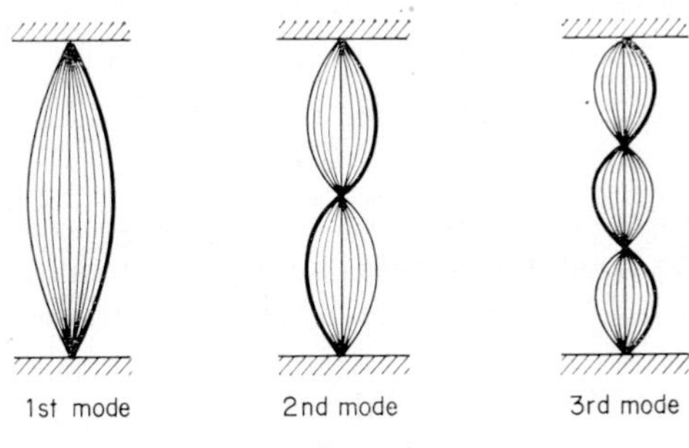

Fig. 2.6

the centre of the string, successive eigenfunctions are alternately cosines and sines with harmonically decreasing wavelengths. Such eigenfunctions are orthogonal to each other, and are easily normalized. All possible eigenfunctions for the string form a countably infinite complete set. Any linear combination of individual eigenfunctions, $a_1u_1 + a_2u_2 + \ldots + a_ju_j + \ldots$, is itself a solution of the whole linear problem. Any pattern of motion whatsoever that the string can display can be represented as a Fourier series, which is merely a special case of Eqn. 2.18, 2.19 above.

Similarly, although with more complicated details, any acoustic pattern that can persist in our simplified ocean can be written as a sum over the normal modes; the r-dependence must be included, of course.

$$\Phi(r, z, t) = Cr^{-1/2} e^{-i\omega t} \mathop{\mathbf{S}}_{j} [A_j k_j^{-1/2} u_j(z)\, e^{ik_j r}] \tag{2.20}$$

† The reason for the increasing number of nodes can be glimpsed from our brief outline of a numerical solution. At the TP's, from Eqn. 2.14, d^2u/dz^2 must vanish; outside the TP's where $V(z) > E$, the allowable solutions $u(z)$ are also restricted by the fact that they must vanish at $z = z_s$ and $z = +\infty$. Nevertheless, $u(z)$ between the TP's must always join, and with continuous first derivative, onto these external solutions. Once some one u_n has been found, the only way that a new u_{n+1} can achieve this join is to go through another half-cycle inside the TP's and then approach a join very much as u_n did.

The complex constant C contains $\exp(-i\pi/4)$; it also provides for including the strength of the source. As in Eqn. 2.18, j denotes n or E as necessary, and **S** is correspondingly a sum or an integral.

2.7 Introduction of acoustic source

We now introduce the assumed point source at $r=0$, $z=z_0$. This source is a mathematical singularity; very near it, the acoustic field is dominated by the source alone, as if it were in an unbounded medium. Hence Eqn. 2.20 must reduce to the very near field of the source as $r\to 0$ and $z\to z_0$. In this procedure, the r-part of Φ must be the Hankel function, Eqn. 2.8, and not the approximate form in Eqn. 2.9. The answer is known; apart from any changes that may occur in the constant C, the result in Eqn. 2.20 is

$$A_j = u_j(z_0) \tag{2.21}$$

We are considering only propagation to long ranges, and so Eqn. 2.20 can be simplified a good deal. It turns out that the totality of 'continuous' modes (those for $E > V_b$ in Fig. 2.5) falls off quite fast with r, because of mutual interferences among the slightly differing adjacent u_E's. For ranges r exceeding a few water-depths H, the total contribution of these continuous modes to Eqn. 2.20 is negligible†. From this fact (not proved here, of course) and Eqn. 2.21, it follows that the acoustic far field in the ocean, caused by a single-frequency c.w. point source at $(0, z_0)$, is given by

$$\Phi(r, z, t) = \phi(r, z)\, e^{-i\omega t} \sim C r^{-1/2} \sum_{n=0}^{N} [k_n^{-1/2} u_n(z_0)\, u_n(z)\, e^{ik_n r}] \tag{2.22}$$

i.e., a sum over the finite number of discrete modes alone. The total number $N+1$ of these modes turns out (from numerical calculations based on data sketched in Fig. 2.3) to be roughly $2{\cdot}4f$, where f is the acoustic frequency in Hz. Hence, e.g., at 100 Hz there are about 240 terms in the sum (Eqn. 2.22), which should be enough to describe quite reliably all but the most complicated of acoustic patterns (the point source, a singularity, is one of the latter exceptions—hence the need for retaining the u_E's up to a certain stage of the analysis).

This is perhaps the time at which to comment on the problem of more

† There is no escaping from conservation of energy. The totality of the continuous modes represents sound rays or waves incident so steeply on the bottom as to fail of total reflection. After a few 'bottom-bounces', practically all of this acoustic energy is lost into the bottom. The sound field remaining thereafter is trapped between perfect reflection at the sea surface and total internal reflection at the bottom interface, and so falls off as $r^{-1/2}$ in amplitude, but with a certain multiplying factor that would have been larger if the bottom had been perfectly reflecting for all angles of incidence.

complicated sources. In general, the near field of a directive source would be most easily expressed in spherical coordinates—a radial distance, a colatitude angle measured from the z axis, and an azimuthal angle in the horizontal plane. The normal modes of the ocean are more conveniently described in cylindrical coordinates, as above, but the azimuthal angle must be included, from Eqn. 2.2 onwards. Fitting the normal-mode sum to the near field of the source would now be much more difficult, although not necessarily impossible. If the main lobe of the source field is directed nearly horizontally rather than vertically, one consequence would be to decrease the role of the continuous modes, because most of the sound would be projected so as to undergo total reflection at the bottom.

At the low frequencies of interest here, such refinements may often be unnecessary. The source is not likely to be very highly directive. If its main lobe is wide enough to insonify the whole depth of the ocean at horizontal ranges of but a few water-depths, the far sound field will still be much like that described for a point source, except that the effective strength of the source will vary with azimuthal angle.

If the source is not c.w. but is pulsed in some way, and has a spread of frequencies, its frequency spectrum can be found by standard methods of Fourier analysis. Then each single frequency in the spectrum behaves as in our description, above, and the total far field can be found by summing, or integrating over, the contributions of all single frequencies. In general, this is more easily said than done, but some cases have been worked out[13, 14]. Moreover, with a source of this kind (e.g., an explosive source) the receiver can be filtered to accept only a fairly narrow band of frequencies, and the received signal can then be time-averaged. There is reasonably good experimental evidence that a signal thus processed behaves much like that from a c.w. source, set to the mid-frequency of the received band.

Eqn. 2.22 is now established as a good representation of the far acoustic field from a point c.w. source, in the simplified ocean. It is also a useful first step in the solution for a more complex source.

2.8 Examination of the mode sum

In principle, we know all the eigenfunctions $u_n(z)$ and their associated eigenvalues $E_n < V_b$. (Also, we have decided that the 'continuous' u_E's with $E > V_b$ need not be considered further.) Each E_n yields a propagation constant k_n (Eqn. 2.13); only positive values of k_n are allowed, to represent outgoing waves at great distances. The next task is to examine and simplify the *mode sum* in Eqn. 2.22.

The amplitude factor $k_n^{-1/2}$ in the mode sum varies little with n. This can be seen from Eqn. 2.13, which show that

$$k_n^2 = \kappa^2 - E_n = (\omega^2/c_0^2) - E_n \tag{2.23}$$

But, also from Eqn. 2.13, V is bounded thus

$$0 \leqslant V(z) \leqslant V_b \tag{2.24}$$

since no E_n exceeding V_b is to be considered. Therefore all E_n's lie between 0 and V_b, and Eqn. 2.23, 2.24 show that

$$\omega/c_b < k_n < \omega/c_0 \tag{2.25}$$

Experimental data suggest that c_b exceeds c_0 by no more than 10 per cent; k_n varies by about this same amount as n changes from 0 to N; and $k_n^{1/2}$ varies by only 2 or 3 per cent about its average value.

Therefore an average value $\langle k_n^{-1/2} \rangle$ can safely be factored out of the mode sum and incorporated in a new C' containing $r^{-1/2}$

$$\phi(r, z) \approx C' \sum_{n=0}^{N} [u_n(z_0)\, u_n(z)\, e^{ik_n r}] \tag{2.26}$$

Next we consider the magnitude of any one term in the mode sum. The exponential is of magnitude unity, whereas $u_n(z)$ varies between some maximum (positive) and some minimum (negative) value. The average value of $u_n^2(z)$, $\langle u_n^2 \rangle$, can be estimated from Eqn. 2.17a. To a first approximation, the integration need be carried out only from z_n', the location of the left-hand TP in Fig. 2.5, to z_n'', the location of the right-hand TP, for outside the TP's the eigenfunctions fall off quite rapidly. Then Eqn. 2.17a becomes

$$\int_{z_n'}^{z_n''} u_n^2(z)\, dz \approx 1 \approx \langle u_n^2 \rangle (z_n'' - z_n') \tag{2.27}$$

Therefore the average maxima and minima of u_n somewhat exceed, in magnitude, $(z_n'' - z_n')^{-1/2}$. For fixed n, $u_n(z)$ can be anywhere between $\pm\alpha(z_n'' - z_n')^{-1/2}$, where α has order of magnitude unity. Because all eigenfunctions have different numbers of nodes, at fixed z $u_n(z)$ has almost randomly scattered values as n runs from 0 to N, provided that N is large. The same behaviour holds for $u_n(z_0)$ as n changes.

Evidently the product $u_n(z_0)\, u_n(z)$ in Eqn. 2.26 is itself nearly a random variable, positive or negative, large or small as n changes with z and z_0 fixed, or as z or z_0 changes with n fixed—but there is one exception. If $z = z_0$, the product becomes u_n^2, which is non-negative although still nearly random in magnitude. For any one n, proper choice of source depth z_0 and receiver depth z can make $u_n(z_0)\, u_n(z)$ assume a maximum magnitude, but in general the values of z_0 and z required to produce the same effect at some other n' will be different.†

The mode sum in Eqn. 2.26 has one more factor in each term, the exponential phase quantity $\exp(ik_n r)$. We have seen that k_n varies by only a few per cent over the whole range of n, and the same must be true of $k_n r$. But at long ranges $k_n r$ is very large, for k_n is approximately equal to κ (Eqn. 2.13), which is ω/c_0 or $2\pi/\lambda_0$, λ_0 being the acoustic wavelength at the minimum sound speed. Since r was specified to be very much larger than the acoustic wavelength, it is clear that always

$$k_n r \gg 2\pi \tag{2.28}$$

Hence a fractional change in $k_n r$ of but a few per cent can make an absolute change as great as or greater than 2π. Consequently the phase term $\exp(ik_n r)$ will vary rapidly as r changes at fixed mode number n, and as n changes at fixed range r. It is worth noting that changes in phase, i.e., in $k_n r$, are completely independent of changes in amplitude caused by altering z or z_0.

We can now see fairly clearly how the mode sum usually behaves. For arbitrary fixed values of z_0, z and r (source depth, receiver depth and range) $\phi(r, z)$ is given by the sum of a great many terms with (usually) almost random amplitudes and phases.

It is then convenient to compute $|\Phi|^2 = \Phi^*\Phi$, where Φ^* is the complex conjugate of Φ; $|\Phi|^2$ is proportional to acoustic intensity. From Eqn. 2.3 and 2.26 we obtain

$$|\Phi|^2 = |C'|^2 \left\{ \sum_{n=0}^{N} u_n^2(z_0)\, u_n^2(z) + \sum_{m,\, n \neq m}^{N}\sum u_n(z_0)\, u_m(z_0)\, u_n(z)\, u_m(z)\, e^{i(k_n - k_m)r} \right\} \tag{2.29}$$

in which $|C'|^2$ is proportional to r^{-1}. Also, via the term $k_n^{-1/2}$ in Eqn. 2.22 (of which an average value was factored out to get Eqn. 2.26), $|C'|^2$ is inversely proportional to acoustic frequency. The single sum over n, in { }, contains only non-negative terms, most of which will not be zero, and this term represents an average acoustic intensity which, if N is large, will

† There is an analogy with the stretched string of Fig. 2.6. Striking it at any one point along its length will stimulate some modes of vibration more than others; e.g., the second mode in the Fig. will certainly not be stimulated by a blow at the middle of the string, whereas the first will. In Eqn. 2.26, $u_n(z_0)$ similarly determines the degree of stimulation of the nth mode, by the source placed at depth z_0. A blind and deaf observer trying to study the vibrations by touching the string will learn nothing about the second mode, but something about the first, third, etc., if he touches it at the centre. In Eqn. 2.26, $u_n(z)$ plays an analogous role in determining how strongly the receiver, at depth z, is insonified.

probably vary little with z and z_0. Separate terms in the double sum, however, contain all the near-randomness of sign, magnitude and phase that we have just been discussing. If N is large, this sum will usually remain near zero, varying somewhat with z, z_0 and r (in the phase term); yet occasionally (e.g., for certain values of r with z and z_0 fixed, etc.) it may be quite large. Hence we expect $|\Phi|^2$ and the acoustic intensity to be governed mainly by $|C'|^2$ times the single sum in Eqn. 2.29, with ordinarily small but occasionally large spatial fluctuations about this value, caused by the double sum. It is obvious that $|\Phi|^2$ is non-negative, always.

The seeming dependence of the single sum on the total number of modes, N, can be analysed approximately. Given a specific profile of sound speed, there are only two ways in which N can be changed. First, an increase in acoustic frequency increases N proportionally (see just below Eqn. 2.22); but $|C'|^2$ is inversely proportional to frequency. Thus the increase in the number of terms in the sum is offset and fairly well nullified, as would be expected on physical grounds. The other way to increase N is to increase the ocean depth H, i.e., to go somewhere else. N is nearly proportional to depth H, and once more it might seem that the single sum, containing only non-negative terms, will increase as N does. We must not, however, forget the normalized property of the eigenfunctions u_n. When the depth H is increased, the added eigenfunctions correspond mainly to eigenvalues $E_n > V_B$ in Fig. 2.5. For E_n well below V_B, the eigenfunctions fall off very fast—exponentially or faster—to the right of their right-hand TP's, and so they are little affected by any moderate change in H. For E_n exceeding V_B, the TP's occur at the sea surface and at the bottom interface, so that $(z''_n - z'_n)$ in Eqn. 2.27 equals H. Therefore for all eigenfunctions corresponding to $E_n > V_B$—not just for the additional ones—Eqn. 2.27 shows that

$$\langle u_n^2 \rangle \approx H^{-1} \tag{2.30}$$

There being two factors of u_n^2 in each term of the single sum, it follows that the 'upper' terms, those for $E_n > V_B$, are multiplied by H^{-2}. Thus the effect of adding more modes, with increased depth, is counterbalanced.

We have just given a semi-mathematical discussion. Physically, there are two effects as the sea depth is increased. First, the impedance seen by the source may change; the result can be of some importance in shallow-water propagation, but it is probably negligible when a great many modes are present. Second, when the depth increases the same source insonifies a greater volume of water; therefore, on the average, the acoustic intensity decreases.

These comments depict only broad averages; changing acoustic frequency or water depth will of course alter the acoustic intensity at some depths and some ranges, and sometimes quite markedly.

'Shallow-water propagation' is properly defined not by water depth alone, but by the fact that the ratio of water depth H to acoustic wavelength λ is not very large compared with unity. Although bottom properties affect the statement somewhat, it is roughly true in shallow water that

$$N \approx H/\lambda \tag{2.31}$$

Then N is not a very large integer, and the randomizing effects in the double sum of Eqn. 2.29 may not be very effective in reducing the value of that sum. Indeed, the sum may vary strongly and fairly regularly with changes in r, producing an effect known as *mode interference*[15].

2.9 Acoustic focusing in the ocean

The discussion just above emphasized the nearly random relations among successive terms of the sum in Eqn. 2.26. An important exception can exist in theory and, as it turns out, in fact also. Suppose that for a substantial number of successive n's in Eqn. 2.26 the corresponding values of k_n change by nearly constant increments

$$k_{n+1} - k_n \approx M \tag{2.32}$$

Then, with M a negative constant, there exists a smallest range, $r = R$, such that $(k_{n+1} R)$ differs from $(k_n R)$ by approximately $\pm 2\pi$. The requirement, in addition to Eqn. 2.32, is that

$$|M| R \approx 2\pi \tag{2.33}$$

Under these conditions, successive values of the phase factor $\exp(ik_nR)$ would scarcely differ, because, of course, $\exp[i(\theta \pm 2\pi)]$ equals $\exp(i\theta)$. This particular subset of modes would not add incoherently, by intensities, as in Eqn. 2.29, but by amplitudes before being complex-squared to get intensity. There would consequently be a potentiality of sound focusing or convergence at range R. If, in addition, source and receiver were at equal depths ($z = z_0$ in Eqn. 2.26), all the amplitude factors would be non-negative, and any focusing would be enhanced.

If focusing exists at range R, there would be a similar tendency at $2R$, $3R$, etc., the phase differences between successive terms being about 4π, 6π, etc., instead of 2π. In general, Eqn. 2.32 is not satisfied exactly for any more than two terms, but it can be nearly true for quite a number. Then the equation can be written

$$k_{n+1} - k_n = M + \epsilon_n, \quad |\epsilon_n| \ll |M| \tag{2.34}$$

and, for ranges that are integral multiples νR of the basic R, Eqn. 2.33 becomes

$$\nu MR + \nu\epsilon_n R = -2\pi\nu + \nu\epsilon_n R \tag{2.35}$$

Evidently the deviation from an exact phase difference of $2\pi\nu$ increases as ν does. This discrepancy must be measured against 2π, not $2\pi\nu$, and it follows that, if sound focusing occurs at range R, it will be progressively poorer at $2R$, $3R$, and so on.

Focusing does not occur only at an exact point, but over some ranges near the value R. Accordingly the problem of focusing needs, and has had, much more thorough investigation than this outline gives. Two different examples exist in the ocean; both were studied by ray methods before wave theory was applied. We shall say more about these cases, later.

2.10 Classification of modes and shapes of eigenfunctions $u_n(z)$

We saw earlier that the eigenfunctions $u_n(z)$ with the corresponding eigenvalues E_n (and therefore, from Eqn. 2.13, values of k_n) can be computed from Eqn. 2.14 with its associated boundary conditions. Computer solutions are straightforward, even for quite large numbers of modes[16]. On the other hand, fairly good analytic or semi-analytic solutions are possible, notably by the Wentzel–Kramers–Brillouin (WKB) approximation[11, 17]. A full computer solution would naturally be programmed to find the acoustic field at (r, z) directly, the various eigenfunctions being calculated, used and discarded *en route* to an answer. Still, there is some point in sketching a few typical eigenfunctions: first, to clarify the whole normal-mode method, and second, to see the possibility of short-cuts.

Fig. 2.7 shows in light solid line the curve of Fig. 2.5, but twisted around to resemble Fig. 2.3 in orientation. The vertical dotted lines separate four different regions. In the first region, labelled $R'R'$, the eigenfunctions are largest near the depth of minimum sound speed, and beyond their respective TP's they fast drop to negligible size. Two normalized eigenfunctions are sketched in heavy solid line, for $R'R'$; their relative amplitudes are roughly correct, and they are plotted at positions corresponding to the associated values of E_n. An acoustic frequency of 100 Hz has been used for the calculation. Evidently these modes are of great importance in discussing the sound field near the SOFAR axis; in complement, they can be strongly stimulated only if the source is in this same range of depths. On the other hand, these modes are of minor or even negligible significance in computing the sound field well above or below their TP's. Indeed, their contributions to the sound field correspond to that of rays that are refracted both above and below the SOFAR axis

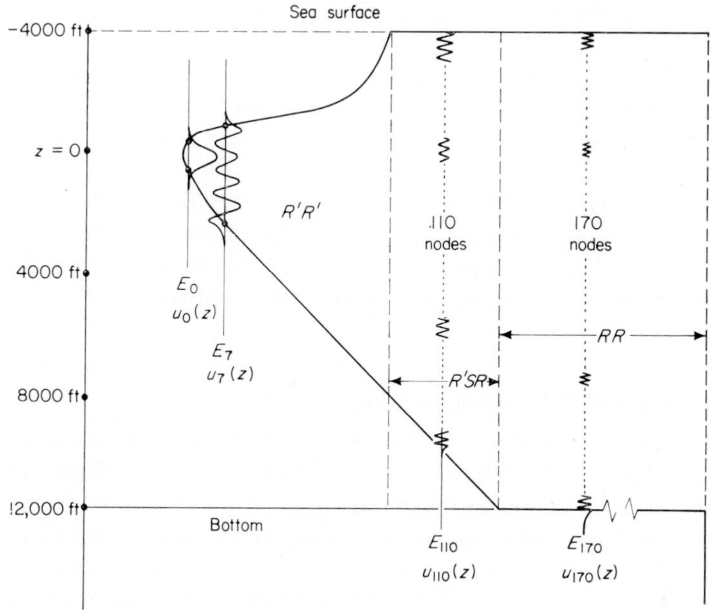

Fig. 2.7

[Fig. 2.4(c)]. The designation $R'R'$ stands for 'refracted, refracted', to mark this correspondence.

In the next region, labelled $R'SR$ (for 'refracted, surface-reflected'), one representative eigenfunction has been drawn, still to about the same scale as the others in $R'R'$. The average variation from minimum to maximum in $R'SR$ eigenfunctions is smaller than in $R'R'$, as a consequence of normalization (see Eqn. 2.27). Evidently $R'SR$ modes can be strongly stimulated and detected fairly near the sea surface (*all* eigenfunctions vanish at the surface, as we saw earlier), but near the bottom they are extremely small in amplitude and so could be neither appreciably stimulated by a source nor appreciably detected by a receiver, there. The contribution of these modes to the sound field corresponds to that of rays reflected upon reaching the sea surface but always refracted up again before reaching the bottom.

The RR modes ('reflected, reflected'—at surface and bottom), of which one sample is sketched, are the only ones in Eqn. 2.26 and 2.29 that are appreciably stimulated by a source, and detected by a receiver, near the bottom. Their exponential fall-offs occur only in the bottom. The discontinuity upon crossing the bottom interface has already been mentioned (p. 36, property 4 of eigenfunctions).

Finally, to the right of region *RR* lies that of the continuous eigenfunctions, described earlier and then disregarded because they die off fast with increasing range. Sketching any single one of these functions is more misleading than helpful, for there is a continuous infinity of them and they interfere mutually to produce a sound field little resembling any one mode of the set.

Two other aspects of the various normal modes are not shown in these sketches: their attenuations with increasing range, and the slightly different speeds with which portions of the sound field corresponding to different modes travel through the ocean. We return briefly to these topics, hereafter.

2.11 *R'R'* and *R'SR* sound focusing

Calculations on *R'R'* modes[11] and on *R'SR* modes[18], both at frequencies in the region of 100 Hz, have shown that Eqn. 2.32 and 2.33 are approximately satisfied for the few lowest-numbered *R'R* modes and for all of the *R'SR* ones. Hence some degree of focusing is expected. In the *R'R'* case, the effect is rather small and probably hopelessly beyond detection, for various reasons[11]; the pertinent range R (Eqn. 2.32) is about 13 nautical miles. For *R'SR*, the effect is well known experimentally, particularly when both source and receiver are near the surface (*surface zone convergence*[19]); the focusing range for a sound speed profile typical of the North Atlantic is about 35 n.m. Ray theory, wave theory and experiment are in fairly good agreement.

Changes in the sound speed profile, within limits, do not alter these effects except to vary the fundamental focal distance R. In particular, it can be shown for *R'SR* focusing that the necessary condition is a profile generally resembling that of Fig. 2.3, provided that the speed of sound, in the water, at the bottom interface exceeds that at the surface.

The upper modes in *R'R'*, i.e., those with larger indices n, and all the *RR* modes fail to show any focusing at all. Since neither *R'R'* nor *R'SR* eigenfunctions are of significance near the bottom, no focusing can be expected from a source placed there, as long as the speed of sound near the bottom exceeds that near the surface.

2.12 Surface-duct modes

Fig. 2.5 is drawn to correspond with a sound speed $c(z)$ that increases monotonically from its minimum, c_m, to some larger value c_s at the surface. It can happen instead that in the upper hundred or few hundred feet the slope of $c(z)$ reverses. Fig. 2.4(b) is an example of this behaviour;

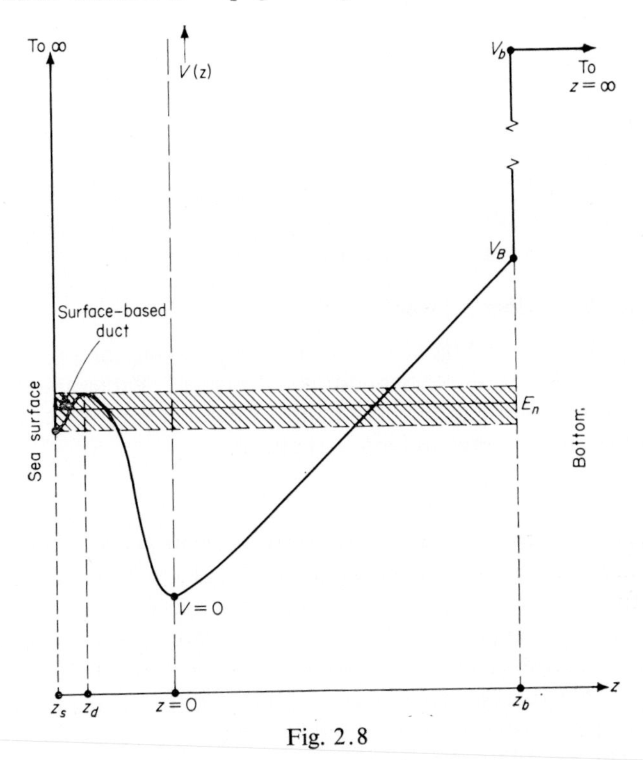

Fig. 2.8

Fig. 2.8 shows the changed Fig. 2.5, with a *surface-based duct* or surface channel between z_s and z_d.

The general effect on our previous analysis is only a change of details, with slight alterations in the values of E_n, k_n and $u_n(z)$. However, for values E_n that cross this jog—the shaded band in Fig. 2.8—there exist eigenfunctions that are of quite large amplitude between z_s and the other TP in the jog, and quite small elsewhere†. These modes are strongly stimulated by a source in the duct and strongly detected by a receiver there.

Accordingly, with source and receiver so placed, the full mode sum of Eqn. 2.26 is dominated by the relatively few terms corresponding to eigenvalues in the shaded band. A good approximation is obtained by neglecting all the other mode terms in the sum, and by greatly simplifying the shape of $V(z)$ to the right of the jog—since it scarcely affects the pertinent u_n's, in any event. The consequence is to single out a certain number of modes, the exact number depending on details of the jog and

† This behaviour is most readily seen by examining the WKB approximation[17] to eigenfunctions.

on acoustic frequency. The eigenvalues for these modes acquire imaginary parts after the rest of $V(z)$ is simplified. These imaginary parts correspond to a slow leakage of acoustic energy out of the modes nearly trapped in the duct, and down into the rest of the ocean.

This special case of acoustic propagation has a counterpart in electromagnetic theory. The problem has been repeatedly analysed for the past 20 years, with increasing detail; both ray[4] and wave[20] methods have been used. Comparisons with experiment are often very good.

2.13 The not-so-spherical cow

In conclusion, something can be said about relaxing the severe restrictions applied to the real ocean conditions of Fig. 2.1. Paragraph numbers, below, agree with those in the list of approximations given after Fig. 2.2, although the order of presentation is changed.

3 *Changes in sea depth H*

Provided that any change ΔH in depth is small per wavelength, i.e., the bottom slope is small, it appears safe to assume that the normal-mode solution for constant depth is only slightly altered, from one range r to another. There is supporting evidence from shallow-water model tests[21]. Specifically, several changes are to be made in Eqn. 2.26 and 2.29. (a) The eigenfunctions $u_n(z_0)$ should be computed for the depth H_0 at the location of the source; (b) $u_n(z)$ should correspond to H at the receiver; (c) N should be set by the minimum depth anywhere along the range. (d) In Eqn. 2.26 the quantity $k_n r$ should be replaced by

$$\int_0^r k_n(r')\,dr'$$

because k_n changes with H, which in turn changes with r. (e) A similar change should be made in any attenuative factor $\exp(-\alpha_n r)$, if α_n changes with H (as it does, rather sensitively, in certain shallow-water cases[21]).

Items (a) and (b) imply that each mode adjusts itself smoothly to slow alterations in sea depth. Item (c) means that one or more of the highest *RR* modes will be cut off entirely at any shallow spot along the transmission path. The energy in the modes thus removed is lost in the bottom[21]. Conversely, the existence along the range of a depth exceeding those at source and receiver allows a larger number, N, of modes, which could not exist at the source and so were not stimulated, and in any event would be cut off again before reaching the receiver. It is doubtful that the extra modes are stimulated unless there are marked roughnesses of surface or bottom interface, or strong inhomogeneities in the water itself.

Since these comments apply only to *RR* modes and for modest changes in depth, propagation governed mainly by *R′R′* or *R′SR* modes should be little affected by changes in depth. If, however, *H* decreases sufficiently, the upper *R′SR* modes will change their character, some becoming *RR* and others 'bottom-reflected, surface-refracted'. Both kinds of altered modes remain oscillatory in shape, and fairly strong, to the bottom. Long range transmission will be relatively worsened, because of bottom losses, yet there will be partial compensation because the source now insonifies a smaller volume.

6a *Horizontal variations in sound speed*

The matter has not been much studied except in ray theory, but probably the normal-mode solution adjusts itself fairly placidly to slow changes, along range *r*, of vertical speed profiles (Fig. 2.3). Ray theory predicts, and wave theory would certainly agree (except perhaps in details), that horizontal variations in sound speed *across* the line between source and receiver cause refraction of sound in horizontal planes. Unless the variations are severe, the effect with a point source would probably be slight, but with a directive source (or directive receiver) there might be substantial deviations from the simpler theory outlined earlier.

Because the mode propagation constant k_n changes with changing depth *H* as well as with changing sound speed, a horizontal pseudo-refraction should also occur when the line from source to receiver crosses a localized slope of the bottom, instead of more or less paralleling it.

5a *Absorption of sound in water*

Down to perhaps 1000 Hz (a very rough dividing line) the amplitude absorption coefficient α in sea water is probably fairly well known[2] as a function of acoustic frequency. A first correction to the mode sum, Eqn. 2.26, would therefore be to insert a single factor $\exp(-\alpha r)$ in the right-hand side, for frequencies around and above 1000 Hz. This is not a perfect correction, as can be seen from a ray picture (e.g. Fig. 2.4). The sound travels from source to receiver by a variety of paths that have quite different lengths. Still, the difference in path lengths is exaggerated in Fig. 2.4, as in most similar diagrams, because the vertical scale of length is greatly enlarged to make perceptible the curvatures of rays.

At frequencies of a few hundred Hz and lower, the true absorption coefficient in the water itself certainly becomes very small. Acoustic measurements over very long travel paths and with very powerful (explosive) sources are necessary if the geometrical attenuation—due to nearly cylindrical spreading—is to be separated from other effects. Therefore all that can really be claimed is that the acoustic *attenuation*, not *absorption*, in excess of geometrical spreading losses, is measured at low frequencies. This extra attenuation is certainly due in part to true absorption, but losses caused by scattering out of the main acoustic path, losses in the bottom, and perhaps non-linear effects, also play parts. In a general way, measurements of attenuation by several observers cohere down to perhaps 50 Hz; yet there are suggestions in the evidence that results differ in widely different parts of the ocean.

Except for this disagreement, in a practical sense the problem is fairly well resolved for very long-range transmission, which can be achieved only by the same methods used to measure the empirical attenuation.

7a *Absorption in and attenuation due to the bottom material*

There is amply convincing although mainly indirect evidence that acoustic losses in the bottom considerably exceed those in the water. Physical mechanisms include direct absorption of longitudinal acoustic waves in the bottom material and, at least in some materials, partial conversion of longitudinal to transverse (shear) waves, with subsequent loss. The latter effect is probably more important in shallow-water cases[6]. Because, as we have seen, $R'R'$ and $R'SR$ normal modes are extremely weak at and in the bottom, these modes are probably very little affected, unless perhaps at extremely great ranges, over which small local losses might accumulate significantly.

In consequence of bottom losses, the totality of continuous modes, which dominate the field near the source, will fall off with range even a little faster than in an idealized case. The main effect, however, is on RR modes, which in ideal theory can propagate to great distances (total internal reflection occurs at the bottom) and yet are also fairly strong at and in the upper regions of the bottom. The totality of these modes will be markedly attenuated at medium and long ranges, to an extent governed by the nature of the bottom, the acoustic frequency and the relative strengths of the individual RR modes.

In shallow-water cases, for which all of the rather few modes are RR, theory can be worked out for absorptive and shear-supporting bottoms—still of simple structure—and model tests agree with theory fairly well[6, 21]. Real bottoms are much more complicated. All that can be said in general is that there are attenuations of the water-borne signal ascribable to the bottom and that, as in theory and model tests, the real-life attenuation coefficients vary from one mode to another[16]. The corresponding ray explanation is that rays striking the bottom at different angles encounter different reflection coefficients.

For deep-water propagation, our knowledge is still more fragmentary. It is harder to determine the physical properties of the bottom anywhere, yet such knowledge is needed for fairly long ranges, over which the physical properties certainly vary considerably. In consequence, the problem has had rather little theoretical attention.

An example of a question that might be open to analysis is the following. Over the very long ranges needed to measure attenuations at low frequencies[2], the main acoustic signal is carried by $R'R'$ modes in the SOFAR channel, where the source and receivers are placed. These modes are exponentially weak at the bottom and so are relatively little affected by bottom losses. In contrast, the accompanying RR modes are weak near the axis, but are much more strongly bottom-attenuated. Are these RR modes completely negligible in such experiments, as factors in producing measurable attenuation? The non-geometrical attenuation is small, and might conceivably be either a large loss in a small part (RR) of the signal or a small loss in a large part ($R'R'$)†.

7b Layered bottom‡

It is thoroughly established that bottom materials are layers of different substances—mud, ooze, clay, sand, rock—having unevenly varying thickness and a wide range of acoustic properties. In two cases, particularly strong influences can be exerted upon sound propagation. First, in shallow-water problems the combination of relatively long acoustic wavelength and small depth ensures that the sound field extends appreciably into the bottom, and may even to some extent be trapped and guided along bottom layers. Second, in deep water, signals beamed at and reflected from the bottom can encounter very complicated reflection conditions (some part of which may be due to bottom topography). Thus, for example, a reflected pulse may not only be well or poorly reflected but elongated and distorted.

In the shallow-water example it has been shown (Reference 7, Chapter 4) that with sufficiently detailed acoustic and geophysical data the apparent vagaries of acoustic transmission can be well explained in retrospect. The great remaining difficulties are that very few thorough experiments exist, to give us more general guidance, and that we lack the ability to predict transmission reliably when only approximate geophysical information exists.

In the deep-water case, several investigators have recently made semi-theoretical studies of specific experiments. The procedure is to postulate a

† The main cause may be quite different, e.g., scattering out of the SOFAR channel by inhomogeneities in the water.

‡ The subjects of this Section and the next two are discussed at length in Reference 7, particularly in Chapters 4 and 6. Many other references are also cited there.

model of the layered bottom, with the help of whatever geophysical knowledge can be gained, and then vary its parameters until the observed acoustic results are reasonably well duplicated. It seems prudent to reserve judgment, because so many parameters are invoked : the number of layers and the thickness, sound speed, density and absorption coefficient of each. Even though physically reasonable values are used for all of these parameters (except perhaps absorption coefficients, the least well understood), there remains a possibility that other greatly different models would have done as well or better.

6*b*, 11*a* *Spatial and temporal variations of sound speed in the water*

Another solid fact is that spatial and temporal variations of sound speed exist in both shallow and deep water. Some of these variations can be discussed fairly easily, being either time-independent horizontal changes in sound-speed profile or rather slow temporal changes (e.g., daily or seasonal). Another kind of semi-regular variation is caused by *internal waves* at the interface between two water layers of different temperatures and densities. These waves are physically akin to those on the surface, but are of much longer period and wavelength, and often of considerable amplitude (many feet). The acoustical effects are usually caused by the heaving and falling of water volumes with quite different temperatures and, therefore, sound speeds†. In principle, such variations can be handled theoretically. Some studies have been made, more by ray than by wave methods. Practical shortcomings are our inability to predict the magnitude and duration of these internal waves, and—except over short ranges—to get reliable measurements of the waves, against which to test acoustic theory and data.

The remaining kinds of variation in sound speed are so widely found through the volume of the ocean and are so widely distributed as to size, duration and strength that whatever their physical causes they can only be regarded as random scatterers of sound. It must be admitted that the primary method of normal modes is not well suited to cope with fluctuations of this sort. The mode method depends on an early separation of the acoustic field variable Φ into t-, r-, and z-parts, whereas these fluctuations are not thus separated. Fortunately, the effects on acoustic propagation are usually relatively small at low frequencies, even though they may accumulate significantly over long paths, and so the fluctuations can be regarded as additive perturbations to the main solution. Then the main

† In regions where large amounts of fresh water impinge upon ocean water, differences in salinity are also heaved about by internal waves, with consequent changes of sound speed.

features of the perturbation can be sought with much simpler primary fields—plane, cylindrical or spherical waves—with which a good deal of work has been done.

Again, the reader is referred to Chapter 6 of Reference 7.

2, 7*c*, 11*b* *Surface and bottom roughness*

Although we lump these two kinds of roughness together in this brief discussion, there are obvious differences. First, surface roughness varies with time but bottom roughness does not (although the motion of a vessel on surface waves may cause tossing about of a directive source or receiver, so that different parts of the bottom are insonified, with a temporal result). Second, there are great differences in both horizontal and vertical sizes. Appreciable surface waves range in vertical amplitude from inches to feet, or exceptionally to tens of feet; bottom roughness runs from pebbles on an otherwise smooth interface to intruding sea mounts or ravines.

If as theorists we cheerfully disregard the tossing about of research vessels, their crews and their sound gear, we find two chief effects of roughness: a greater attenuation in the propagation of sound, and temporal or spatial fluctuations in received signals. In terms of normal-mode theory, both effects have the same cause. The acoustic energy carried by any one of the orderly and persisting modes described earlier is continually being scattered by rough surfaces into other modes. That is, a rough patch acts rather like another acoustic source when insonified by the main acoustic field. Thus $R'SR$ modes will be to some extent scattered by surface roughness into RR modes and some of the continuous ones; both kinds suffer bottom losses. RR modes will be partly scattered by either surface or bottom roughness into higher RR modes and continuous ones, which suffer increased bottom losses. $R'R'$ modes are probably little affected by roughness.

It might seem that a balance could be achieved as more lossy modes are simultaneously scattered into less lossy ones, but the preponderance is towards net loss, probably for three main reasons. First, the least lossy modes ($R'R'$) are not stimulated appreciably by a rough surface or bottom, being exponentially weak there. Second, there are many more RR and continuous modes than $R'R'$ and $R'SR$ combined, so that the probabilities incline towards loss and not gain. Third, any acoustic energy that is scattered, part way along the transmission path, into continuous modes suffers a quick and final end, for this part of the sound field dies off fast not far from its point of origin.

That time-varying roughness produces fluctuations in received signals can be seen from Eqn. 2.26 and the discussion that followed it. Eqn. 2.26 presents the sound field at r, z as a sum of terms with (usually) nearly random amplitudes and phases; that is, ϕ is the sum of nearly random vectors. Introducing surface roughness somewhere between source and receiver will not much change the propagation constants k_n, particularly in deep-water problems, but it will considerably reshuffle the amplitudes of the separate vectors. Therefore the sum vector ϕ will generally change in both magnitude and phase and, the surface roughness being time-varying, the resulting changes in ϕ will also vary with time. Spatial fluctuations result if either source or receiver is moved to a new location, an act which changes the roughnesses encountered over the acoustic path and so alters ϕ.

In shallow-water propagation at fairly low frequencies, fluctuations of signal caused (presumably) by strong surface waves may be many decibels in size. Apart from numerous studies of surface scattering *per se* (Ch. 3 of this book), most analyses of propagation with scattering have been made in shallow water. Chapter 6 of Reference 7 is highly pertinent. A 'semi-model' experiment[22] has provided many interesting data on acoustic fluctuations caused by measured surface waves. A recent theoretical study[23], not yet published, successfully explains most of these observations as caused by scattering at the surface, and not by local variations in water depth.

To go much beyond these descriptive comments would require far more space, more mathematics and indeed a good deal of theoretical research not yet essayed. We therefore conclude the discussion of normal-mode theory for propagation of underwater sound, leaving it for the reader to examine the extensive literature and try his own hand at unsolved problems, if he so wishes.

References

(a) Five books written in or (in one case) translated into English deal at considerable length with wave theory of underwater sound. We list them in order of publication dates, with comments.

(i) *Propagation of Sound in the Ocean* by W. M. Ewing, J. L. Worzel and C. L. Pekeris; Memoir 27 of the Geological Society of America (October 1948). Consists of three papers, with repeated (and hence confusing) pagination. The middle paper, about half of the whole Memoir, concerns theory, in the main. It is a classic in terms of priority—at least in the English language—but is not always clear, and it is to some extent repetitive.

(ii) *Elastic Waves in Layered Media* by W. M. Ewing, W. S. Jardetzky and F. Press (McGraw-Hill, New York, 1957). Has a strong flavour of applied mathematics inspired by a class of physical occurrences mainly related to propagation of underwater sound. Not devoid of physical results or presentation of data, yet a difficult book for a beginner in the subject, unless he is already well grounded in the fields of applied mathematics exploited by the authors.

(iii) *Introduction to the Theory of Sound Transmission* by C. B. Officer (McGraw-Hill, New York, 1958). Avowedly a textbook rather than a monograph; therefore most of the mathematical treatment is drawn from other sources (a good deal from (i) above), but without direct citation of the originals to facilitate further study. There is a list of suggested reading after each chapter. The book contains a good deal of useful descriptive material, but the mathematical presentations are isolated, not integrated together. A numher of minor errors, and a few more serious ones, decrease its usefulness.

(iv) *Waves in Layered Media* by L. M. Brekhovskikh; English translation by D. Lieberman, edited by R. T. Beyer, of 1956 Russian version (Academic, New York, 1960). An orderly and powerful mathematical development of the whole subject given in the title, with parallel instances in electromagnetism and in acoustics—most of the latter, in underwater sound. Physical results are presented for comparison with theory, or as starting points. A great many references are cited, often, naturally enough, in the Russian literature. A book well worth owning by anyone interested in the theoretical side of underwater sound; its drawbacks for the beginner are the range of mathematics and the fact that it is not intended only or mainly as a treatment of underwater sound, so that pertinent topics are separated by other material.

(v) *Ocean Acoustics* by I. Tolstoy and C. S. Clay (McGraw-Hill, New York, 1966). By all odds the best book to date for a physical and mathematical (rather than engineering) treatment of underwater sound. It has an ample list of references to other books and to papers; it displays a wide range of experimental results; it covers more ground than do (i) and (iii); and its mathematics—although by no means elementary—remains much less formidable than that in (ii) and (iv). In mathematical approach, the present chapter more closely resembles Tolstoy's and Clay's book than any of the others, although the methods—not the results—still differ somewhat.

(b) References cited are not necessarily the pioneering work in the pertinent topic, and they may not be definitive, or nearly so. Sometimes a specific reference has been chosen because it, in turn, cites much related work; sometimes, because its mathematical treatment resembles that used in this chapter; sometimes, because it opens up or suggests avenues of exploration.

1 See Section (a) of the reference list which concerns books available for study.
2 Thorp, W. H., *J. Acoust. Soc. Am.*, **38**, 648 (1965).
3 Marsh, H. W., R. H. Mellen and W. L. Konrad, *J. Acoust. Soc. Am.*, **38**, 326 (1965).
4 Pedersen, M. A., *J. Acoust. Soc. Am.*, **34**, 1197 (1962).
5 Ewing, W. M., and J. D. Worzel, 'Long Range Sound Transmission', Part 3 of 'Propagation of Sound in the Ocean', *Geol. Soc. Am. Mem.*, **27** (1948).
6 Williams, A. O., Jr., and R. K. Eby, *J. Acoust. Soc. Am.*, **34**, 836 (1962).
7 Tolstoy, I., and C. S. Clay, *Ocean Acoustics* (McGraw-Hill, New York, 1966), pp. 118–123.
8 *Physics of Sound in the Sea*, Part 1, 'Transmission' (Committee on Undersea Warfare). (Originally Div. 6, Vol. 8, U.S. N.D.R.C. Summary Technical Reports.)
9 Lindsay, R. B., and H. Margenau, *Foundations of Physics* (Wiley, New York, 1936), pp. 440 ff.

10 Williams, A. O., Jr., *J. Acoust. Soc. Am.*, **32**, 363 (1960).
11 Williams, A. O., Jr., and W. Horne, *J. Acoust. Soc. Am.*, **41**, 189 (1967).
12 Schiff, L. I., *Quantum Mechanics* (McGraw-Hill, New York, 2nd edition, 1955), pp. 50–52.
13 Pekeris, C. L., *Theory of Propagation of Explosive Sound in Shallow Water*, Part 2 of Mem. 27, Ref. 5.
14 Ref. 7, pp. 105 ff.
15 Ref. 7, pp. 103, 104, 129 ff.
16 Tolstoy, I., and J. May, *J. Acoust. Soc. Am.*, **32**, 655 (1962). See also Ref. 7, Chapter 5.
17 Ref. 12, pp. 184–193.
18 (a) Carter, A. H., *Doctoral Dissertation*, Brown University, Providence, R.I., U.S.A. (1963). (b) Carter, A. H., and A. O. Williams, Jr., *J. Acoust. Soc. Am.*, **34**, 1985(A) (1962).
19 Hale, F. E., *J. Acoust. Soc. Am.*, **33**, 456 (1961).
20 Gordon, D. F., and M. A. Pedersen, *J. Acoust. Soc. Am.*, **35**, 810(A) (1963).
21 Eby, R. K., A. O. Williams, Jr., R. P. Ryan and P. Tamarkin, *J. Acoust. Soc. Am.*, **32**, 88 (1960).
22 Scrimger, J. A., *J. Acoust. Soc. Am.*, **33**, 239 (1961).
23 Halpin, H., *Master's Thesis*, Brown University, Providence, R.I., U.S.A. (1967).

3

Scattering from the Sea Surface

Herman Medwin
Naval Postgraduate School, Monterey, California, U.S.A.

3.1 Introduction

In studying sound propagation in the sea one would like to be able to ignore the complications due to the agitated sea surface. But even in very deep oceans the increase of speed of propagation with increasing depth ultimately refracts rays of sound up towards the surface; in shallow seas, reflection at the sea floor has the same effect at still shorter ranges. Consequently, whether one uses an isotropic source such as an explosive, or a beam, it is clear that the interaction of the sound rays with the ocean surface must be considered if we are to predict surface reverberation and the mean value and variance of the sound pressure at distant regions. From a different point of view, the study of scattering of sound from the sea surface can answer questions about the oceanographic characteristics of the wind-driven interface. Both the radar specialist and the acoustician have probed this problem.

To the uninitiated the incidence of underwater sound on the sea surface appears as one of the simplest problems of classical physics; for is not the surface of the water simply a plane interface between two fluid media? This being the case, why not simply use the Rayleigh equation (Eqn. 3.1) for the ratio of reflected acoustic pressure to incident pressure for a ray moving from a medium of density ρ_1 and speed c_1 to a medium $\rho_2 c_2$ incident at angle θ_1?

$$\frac{p_r}{p_i}=\frac{(\rho_2/\rho_1)-[c_1/(c_2\cos\theta_1)]\sqrt{1-(c_2/c_1)^2\sin^2\theta_1}}{(\rho_2/\rho_1)+[c_1/(c_2\cos\theta_1)]\sqrt{1-(c_2/c_1)^2\sin^2\theta_1}} \tag{3.1}$$

The answer is that Eqn. 3.1 is certainly valid if we have a plane wave incident on a non-viscous, plane water–air face. It then yields the simple and somewhat dull result $p_r=-p_i$ independent of angle of incidence. For a non-plane surface, the equation may be valid on a point-to-point basis if the surface is 'not too rough', a qualification that we must consider very carefully. The equation is certainly not useful in the shadow regions of a very rough surface.

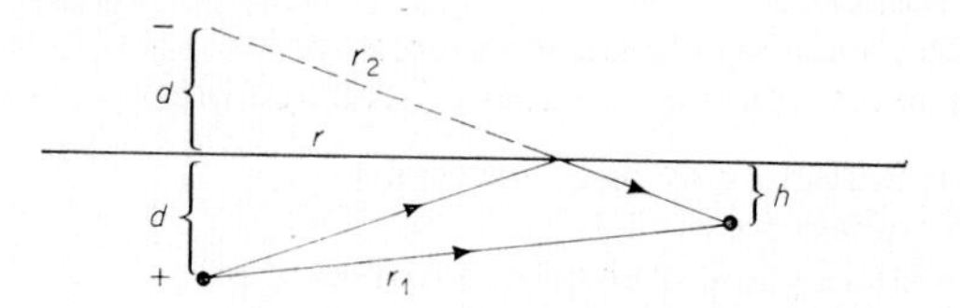

Fig. 3.1 Geometry for Lloyd's Mirror effect

We will confront reality, at once, by considering the Lloyd's Mirror problem as observed at sea[1]. The geometry is shown in Fig. 3.1. If the surface is truly smooth the isotropic continuous wave propagation from a point source can be studied by assuming that the effect of the boundary condition at the surface is to supplement the original pressure field

$$P_1=\frac{B}{r_1}e^{j(\omega t-kr_1)} \tag{3.2}$$

by an out-of-phase image contribution

$$P_2=-\frac{B}{r_2}e^{j(\omega t-kr_2)} \tag{3.3}$$

so that the total field is the sum of the two. For source depth, d, and hydrophone depth, h, both small compared to the horizontal range, r, the interference pattern which results from the coherent surface reflection is given by

$$P=\frac{2jB}{r}\sin\frac{khd}{r}\,e^{j(\omega t-kr)} \tag{3.4}$$

There are peaks and troughs up to the critical range $r_c=4hd/\lambda$. Beyond that range the difference of the distances from the field point to the surface and to its image can no longer be great enough to create the phase shift necessary for additional maxima and minima to occur. For $r>r_c$ there is a rapid decrease in sound pressure (12 dB/double distance).

The early attempt to solve the surface image problem for a rough sea was strictly empirical. It was clear that a rough surface could not reflect perfectly and, therefore, that the reflected sound could neither completely

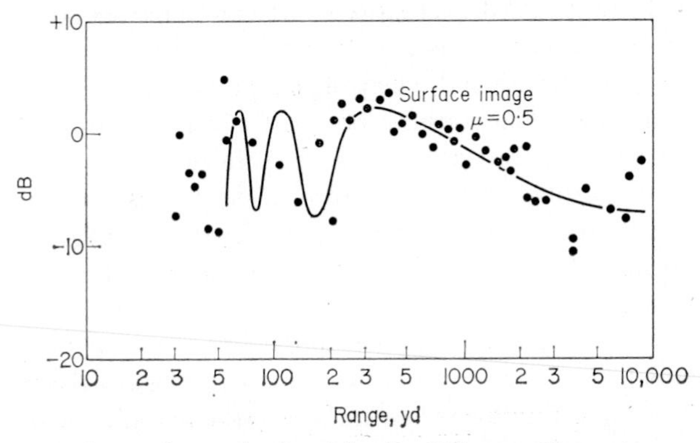

Fig. 3.2 Experimental graph for Lloyd's Mirror effect at sea. Spherical divergence has been subtracted out. A reflection coefficient 0·5 has been used to empirically fit theory to the data. Frequency is 1·8 kHz[1].

cancel the direct sound nor add equally to it to give the +6 dB peaks of the interference pattern. The reflected pressure P_2 was therefore arbitrarily multiplied by a factor μ less than unity, called the 'effective reflection coefficient' and determined by experimental curve fitting. Fig. 3.2 shows the modified theory with the spherical divergence loss (6 dB/double distance) subtracted out. Such an *ad hoc* explanation does not yield any information about the ocean surface. Further, although the optical case may be characterized by such a reflection coefficient, the ocean surface is continually in motion and the factor μ in no way helps to describe the variation of scattered pressure as time goes on. A new approach was needed.

The effect of the roughness of the sea surface is, of course, the core of the problem. To describe the scattering of sound from the sea surface, one must find first a suitable description of the sea surface. This is not necessarily the measurement and description used by the oceanographer. In general, the oceanographer's wave recorders are insensitive to 'capillaries' (of wave length $\Lambda \leqslant 2\cdot 4$ cm). Furthermore, the oceanographer often processes his data particularly to eliminate the longest wave length components. The acoustician concerned with propagation of frequencies of order 60 kHz ($\lambda = 2\cdot 5$ cm) or higher may well find that the presence or absence of capillaries has a significant effect on the scattering of sound, particularly for the back direction.

Another coordinate of the surface scattering problem is the presence of near-surface volume scatterers. Minute bubbles, having resonant acoustic scattering effects of the order of 1000 times greater than that suggested by their geometrical cross-sections, can be particularly significant. The relative importance of underwater volume inhomogeneities becomes greater in back-scattering and when the sound beam is nearer to grazing incidence. One must then consider both the effects of surface scatter and volume scatter from points near the surface. In practice, these two effects often cannot be resolved for two reasons: first, a pulse of finite duration will be scattering from the space near the surface simultaneously with that from the surface itself; second, even if the pulse duration were infinitesimal, when incidence is non-normal one edge of the finite beam will strike the surface while the other parts of the beam (not having yet reached the surface) will be scattering from the volume just below the surface.

We shall consider first the rough surface of the sea, then scattering from the rough surface (Section 3.2). We shall emphasize the experimentally important cases of backscatter including the potential effect of bubbles (Section 3.3) and scatter in the specular direction (Section 3.4). An appendix is provided for the essential bubble theory needed in Section 3.3.

3.2 The wind-driven rough surface and its effect on incident sound

3.2.1 The sea

Perhaps the simplest description of the wind-driven ocean surface is that it is 'chaotic'. Apt though this poetical term is, we must replace it by an appropriate mathematical description if we are to evaluate the effect of the surface on underwater sound. The descriptions that are commonly used depend closely on the measuring devices and techniques of analysis. One of the earliest measurements was the probability density of surface heights at a fixed position. Wave-height data can be taken by any of several instruments such as vertical resistance wires, capacitance probes, bottom mounted pressure recorders, accelerometers, stereocameras and motion pictures of the surface at a vertical standard. Knowledge of the distribution of the surface slopes has been obtained directly by observation of glitter caused by reflection of light from a flash bulb at night[42] and by using sunlight.

One of the most useful experiments, for our purpose, is that of Cox and Munk[2, 3]. Fig. 3.3 shows the distribution of slopes in the upwind–downwind direction as well as the crosswind direction for glitter experiments carried out in the open sea. In another experiment at the same

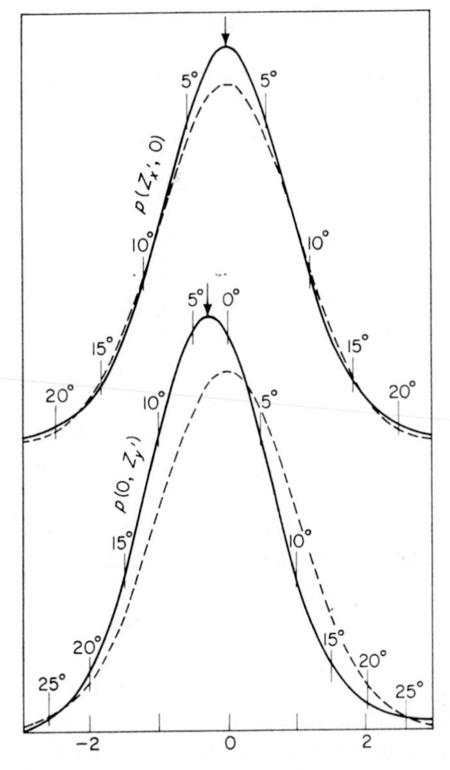

Fig. 3.3 Principal sections through the probability distribution surface $p(Z_x', Z_y')$. The upper curves are along the crosswind axis; the lower curves along the upwind axis. The solid curves refer to the observed distribution, the dashed to a Gaussian distribution of equal mean square slope components. The thin vertical lines show the scale for the standardized slope components $\xi = Z_x'/\Sigma_u$ and $\eta = Z_y'/\Sigma_c$. The heavy vertical segments show the corresponding tilts $\beta = 5°, 10°, \ldots, 25°$ for a wind speed of 10 m s^{-1}; the skewness shown in the lower curve is computed for this wind speed. The modes are marked by arrows[2]

location a mixture of oil was used over an area of one-quarter square mile in order to determine the effect of elimination of high frequency ripples; the oil was found to reduce the mean square slopes by a factor as large as three depending on the wind speed and to eliminate skewness but to leave the peakedness unchanged. The values of mean square slope for the cross-direction, Σ_c^2, and for the up–down wind direction, Σ_u^2, and the mean square slope regardless of direction, $\Sigma_c^2 + \Sigma_u^2$, were found to be linear with wind speed W (m s^{-1} measured 41 ft above the surface):

Clean surface

$$\Sigma_c^2 = 0\cdot003 + 1\cdot92 \times 10^{-3} W \pm 0\cdot002$$

$$\Sigma_u^2 = 0\cdot000 + 3\cdot16 \times 10^{-3} W \pm 0\cdot004$$

$$\Sigma^2 = \Sigma_c^2 + \Sigma_u^2 = 0\cdot003 + 5\cdot12 \times 10^{-3} W \pm 0\cdot004 \tag{3.5}$$

Oil-slick surface

$$\Sigma_c^2 = 0\cdot003 + 0\cdot84 \times 10^{-3} W \pm 0\cdot002$$

$$\Sigma_u^2 = 0\cdot005 + 0\cdot78 \times 10^{-3} W \pm 0\cdot002$$

$$\Sigma^2 = \Sigma_c^2 + \Sigma_u^2 = 0\cdot008 + 1\cdot56 \times 10^{-3} W \pm 0\cdot004 \tag{3.6}$$

The anisotropy of the slope, measured by the ratio Σ_u^2/Σ_c^2, varied from 1·0 to 1·8 depending on the gustiness of the wind.

It is significant that the large deviations from normality (skewness and peakedness) observed in the above slope experiment were not found in data of Kinsman[4] which contained no appreciable energy at frequencies greater than 2·5 Hz. Further, the deviations are virtually undetectable in MacKay's extensive work[5] which was based on data obtained by bottom pressure recorders insensitive to high frequency waves. In both cases, very low frequency waves propagated from distant storms (swell) were eliminated in the data analyses.

Time records of displacement at a point have also been analysed to obtain one-dimensional wave energy spectra, as a function of frequency, in analogy to acoustical spectra. This type of analysis has proven to be particularly fruitful in oceanographic studies of wave growth, attenuation and stability as a function of wind speed and duration[6]. Idealized equilibrium ocean spectra have been defined[7, 8, 9, 43]. There is an excellent monograph on the subject[10].

The relation between the surface mean square slope and the wave frequency spectrum, is

$$\Sigma^2 = \int_0^\infty K^2 G(\Omega)\, \mathrm{d}\Omega \tag{3.7}$$

where Ω = component wave frequency s^{-1}

K = component wave number = $2\pi/\Lambda$

Λ = component wave length

$G(\Omega)$ = component spectral energy density (cm^2 s)

We shall also be interested in the mean square height of the surface, σ^2, which is related to the wave spectrum by

$$\sigma^2 = \int_0^\infty G(\Omega)\,d\Omega \tag{3.8}$$

It has been shown[2, 3] that a sea surface with an arbitrarily wide continuous spectrum of waves produces a Gaussian distribution of surface slopes. A Gaussian distribution of surface heights and a Gaussian surface correlation function imply a Gaussian distribution of surface slopes, provided that the surface is gently undulating[11].

It is clear that high frequency components play a much greater part (through K^2) in the mean square slope than they do in the mean square height of the surface wave structure.

3.2.2 *Scattering solution using the Helmholtz integral*

A theory of sound scattering with application to a Gaussian distribution of slopes, has been derived separately by Isakovitch[12] and Eckart[13] and extended by Beckmann[11] and many others. The method starts with the Helmholtz integral for the field P_i at an interior point of a volume in terms of the total field P_{rs} at the surface, and its *inward* normal component

$$P_2 = \frac{1}{4\pi}\int_S \left[P_{rs}\frac{\partial}{\partial n}\left(\frac{e^{ikr}}{r}\right) - \frac{e^{ikr}}{r}\frac{\partial P_{rs}}{\partial n}\right] dS \tag{3.9}$$

In the 'Kirchhoff method', assumed in this section, the integration is performed by (1) approximating the values of P_{rs} and $\partial P_{rs}/\partial n$ at the scattering surface by the values that would be present on a fully illuminated tangent plane at each point. Since the approximation is best for sound wave lengths much less than the radius of curvature at each position it is essentially a 'high frequency' or 'geometrical optics' assumption and the method is sometimes identified by those names. Additional assumptions are

(2) the interior point at which P_2 is measured is in the far field or 'Fraunhofer Zone' of radiation from the surface,

(3) there is no shadowing by surface elements,

(4) there is no multiple scattering.

In general we follow the particular development and notation of Beckmann[11]. However to compare with experiment in which a piston source is used it is desirable to correct the error inherent in the common assumption of uniform ensonification over a fixed area. The modification is most easily done in the case of rectangular surface ensonification in which it is convenient to replace the true piston field incident at the surface

$$P_1 = P_{axis}\left[\frac{\sin(kB\sin\alpha)}{kB\sin\alpha}\right]\left[\frac{\sin(kD\sin\beta)}{kD\sin\beta}\right] \tag{3.10}$$

by the approximation

$$P_1 = P_{axis}\cos\left(\tfrac{1}{2}u_x x\right)\cos\left(\tfrac{1}{2}u_y y\right) \tag{3.11}$$

To integrate over the surface area we require the angles of the surface position to be approximated by

$$\sin\alpha \simeq \frac{x\cos\theta_1}{r_1} \qquad \sin\beta = y/r_1$$

so that

$$u_x = \frac{kB\cos\theta_1}{r_1} \qquad u_y = kD/r_1$$

where $k = 2\pi/\lambda$

r_1 = distance from source to surface

α, β = angles between ray and beam axis measured in, and perpendicular to, plane of incidence, respectively

B, D = half-length and half-width of the rectangular piston source

θ_1 = angle of incidence, measured between beam axis and normal to the surface.

The limits of the ensonified area are taken to be $u_x X = u_y Y = \pi$ which defines the complete main lobe of the diffracted beam with small error.

For the general wind-agitated surface the expression for the mean relative scattered pressure (compared to smooth surface reflection) is then obtained from the Helmholtz Integral

$$\langle R\rangle = \left\langle\frac{P_2}{P_{20}}\right\rangle = 16F/(u_x u_y)$$

$$\times \int_{-X}^{+X}\int_{-Y}^{+Y} \exp(iv_x x + iv_y y)\cos\left(\frac{u_x x}{2}\right)\cos\left(\frac{u_y y}{2}\right)\langle e^{iv_z\zeta}\rangle\,dx\,dy$$

$$\simeq F\left(\frac{\pi+2}{4}\right)\langle e^{iv_z\zeta}\rangle\,[\mathrm{sinc}\,(v_x + u_x/2)\,X]\,[\mathrm{sinc}\,(v_y + u_y/2)\,Y] \tag{3.12}$$

where

$\zeta(x, y)$ = surface height above the reference plane

X, Y = half-length and half-width of ensonified surface

$$\operatorname{sinc}\gamma = \frac{\sin\gamma}{\gamma}$$

θ_2 = angle of scatter

θ_3 = angle between scatter plane and incidence plane

P_2 = instantaneous scattered pressure for rough surface

P_{20} = scattered pressure for plane surface when $\theta_2 = \theta_1$

$$F = \frac{1 + \cos\theta_1 \cos\theta_2 - \sin\theta_1 \sin\theta_2 \cos\theta_3}{\cos\theta_1 (\cos\theta_1 + \cos\theta_2)}$$

$$v_x = k(\sin\theta_1 - \sin\theta_2 \cos\theta_3)$$

$$v_y = -k \sin\theta_2 \sin\theta_3$$

$$v_z = -k(\cos\theta_1 + \cos\theta_2).$$

We now assume that the surface heights are a stationary random function with mean value zero over the surface at any instant and over time, at any surface position: $\langle\zeta\rangle = 0$. In particular the height distribution is assumed to be 'normal' or 'Gaussian', given by

$$w(z) = \frac{1}{\sigma\sqrt{2\pi}} e^{-z^2/2\sigma^2} \tag{3.13}$$

where σ^2 is the variance of heights about the mean.

For this distribution, the characteristic function is

$$\chi(v_z) \equiv \langle e^{iv_z\zeta}\rangle \equiv \int_{-\infty}^{+\infty} w(z)\, e^{iv_z\zeta}\, dz = e^{-g/2} \tag{3.14}$$

where

$$\sqrt{g} = 2\pi \frac{\sigma}{\lambda} (\cos\theta_1 + \cos\theta_2) \tag{3.15}$$

is the very important 'roughness parameter' for the surface.

We also introduce the correlation of the heights at nearby surface positions

$$C(l) \equiv \frac{\langle\zeta_1\zeta_2\rangle - \langle\zeta_1\rangle^2}{\langle\zeta_1^2\rangle - \langle\zeta_1\rangle^2} \tag{3.16}$$

where $\zeta_1 = \zeta(x_1, y_1)$; $\zeta_2 = \zeta(x_1, y_2)$; $l = |\overrightarrow{(x_1, y_1)} - \overrightarrow{(x_2, y_2)}|$; and in particular assume an isotropic Gaussian form

$$C(l) = e^{-l^2/L^2} \tag{3.17}$$

where L is the correlation distance. Beckmann[11] shows that the mean square slope can be expressed in terms of the two parameters for a Gaussian surface

$$\Sigma^2 = 2\sigma^2/L^2 \tag{3.18}$$

This completes our elementary description of the surface.

Returning to Eqn. 3.10, since

$$R = \frac{P_2}{P_{20}}$$

is complex we seek the mean square value $\langle RR^* \rangle$.

The key to the evaluation of $\langle RR^* \rangle = \langle R \rangle \langle R^* \rangle + D\{R\}$ is the determination of the variance $D\{R\}$. Using our cosine shading function we obtain

$$D\{R\} = (\pi^3 F^2 u_x u_y/32) \int_{l=0}^{\infty} J_0\left(l\sqrt{(v_x+u_x/2)^2+(v_y+u_y/2)^2}\right) [\chi_2 - \chi\chi^*]\, l\, dl \tag{3.19}$$

where $\chi_2 = e^{-g(1-C)}$ is the two-dimensional characteristic function. Since

$$\chi_2 - \chi\chi^* = e^{-g} \sum_{m=1}^{\infty} \frac{g^m}{m!} e^{-ml^2/L^2}$$

for a Gaussian correlation function, we obtain

$$D\{R\} = (\pi^3 F^2 e^{-g} L^2 u_x u_y/64) \sum_{m=1}^{\infty} \frac{g^m}{mm!} \exp\{-[(v_x+u_x/2)^2+(v_y+u_y/2)^2]\, L^2/4m\} \tag{3.20}$$

Finally, identifying $4XY = A$, the area ensonified, the mean square relative scattered pressure is

$$\langle RR^* \rangle = (16F^2 e^{-g}/\pi^4) \left\{ [\mathrm{sinc}^2\,(\tfrac{1}{2}u_x+v_x)\, X]\, [\mathrm{sinc}^2\,(\tfrac{1}{2}u_y+v_y)\, Y] + \frac{\pi L^2}{A} \sum_{m=1}^{\infty} \frac{g^m}{mm!} \exp\{-[(\tfrac{1}{2}u_x+v_x)^2+(\tfrac{1}{2}u_y+v_y)^2]\, L^2/4m\} \right\} \tag{3.21}$$

Before looking in greater detail at special cases of this equation we observe that the first term is significant only for specular scattering ($v_x = v_y = 0$) where it dominates the total scatter for relatively smooth surfaces; this term represents the coherent component of the total scatter. The second term is dominant for scattering in non-specular directions for any sea and in the specular direction, as well, if $g \gg 1$; it defines the randomly-phased contributions from the many randomly-positioned, locally-correlated facets in the ensonified area.

The relative scattered intensity measured at distance r_2 from the scattering area, compared with the intensity of the *incident* plane wave, can be obtained from

$$\left\langle \frac{|p_2|^2}{p_1^2} \right\rangle = \langle RR^* \rangle \left(\frac{A \cos \theta_1}{\lambda r_2} \right)^2 \tag{3.22}$$

In the next two sections of this chapter we shall see how well these predictions fit the experimental data.

3.3 Backscattering

The sound which is scattered back from the ocean surface to the source position ('backscattering' or 'surface reverberation') has plagued the users of underwater acoustics from the very first days of the sonar era. The reviews of studies during and since World War II are burdened with unresolved data about surface reverberation as a function of sound frequency, sea state, pulse duration, angle of incidence and so on and on. With the theoretical developments such as presented in Section 3.2 acousticians finally seemed to have the means for understanding, from physical principles, the essential causes of surface backscatter.

3.3.1 Test of the solution

Almost all of the data available[14, 17] are for scattering in the realm $g > 1$ and so we proceed directly to the case of rough surface backscatter by setting $g > 1$ and $\theta_2 = -\theta_1$ in our theory. Furthermore, because source dimensions and beam pattern are rarely stated in the literature of surface scattering experiments we re-frame our equation in terms of a uniform beam intensity by allowing $u_x X \simeq u_y Y \to 0$.

Finally, for direct comparison with experiment we define the backscattering coefficient for the surface

$$S_s = \left\langle \frac{|P_2|^2}{|P_1|^2} \right\rangle \Big/ (A/r_1^2) \tag{3.23}$$

We then have†

$$S_s = (8\pi\Sigma^2 \cos^4 \theta_1)^{-1} \exp(-\tan^2 \theta_1 / 2\Sigma^2) \tag{3.24}$$

The relation between wind speed and mean square slope (Eqn. 3.5) leads immediately to curves such as shown to the right of Fig. 3.4. Two speeds are selected to bracket the 8–10 knot wind case reported by Urick and Hoover[15, 25]; the experimental scattering levels are indicated by

† Eckart's derivation, which assumes an anisotropic surface, can be written as

$$S_s = (8\pi\Sigma_c\Sigma_u)^{-1} \exp\left\{-\frac{1}{2}\left[\left(\frac{\cos\theta_{1x}}{\Sigma_u \cos\theta_1}\right)^2 + \left(\frac{\sin\theta_{1x}}{\Sigma_c \cos\theta_1}\right)^2\right]\right\}$$

where θ_{1x} = angle between the plane of the wind (the xz plane) and the plane of sound incidence.

circles. The agreement between theory and experiment is reasonable only from normal to 30° incidence. The height and location of the anemometer are unspecified. Chapman and Scott[20] show similar agreement between theory and experiment again only near normal incidence for octave bands from 0·2 to 6·4 kHz.

These results are also consonant with the 60 kHz experiment[18] in which the authors demonstrate the sensitivity of sound scatter to onset time of winds and comment on the importance of getting accurate wind speeds at the location and precise time of the experiment. Unfortunately, in no case was the distribution of sea heights or slopes determined at the time

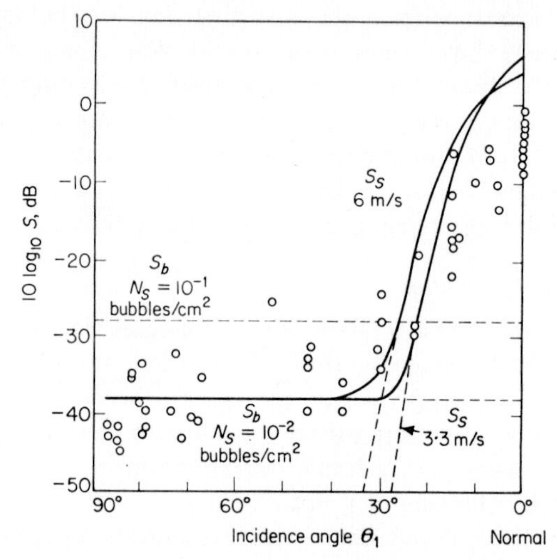

Fig. 3.4 Experimental and theoretical surface-scattering factors. The scattering factors are shown for the surface-scattering factor S_s and the bubble scattering factor S_b. The data are Urick and Hoover[15], 8–10 knot wind or 4–5 m s^{-1}. (From Clay and Medwin[16])

and place of the scattering experiment and so belated application of Eqn. 3.5 is required in order to compare with theory. The ocean experimental data are summarized in Fig. 3.5[14].

Controlled sound scattering experiments have been conducted[21, 22] in the laboratory simultaneously with measurements of the wind-agitated surface of a large anechoic tank. The surface-wave characteristics were analysed by a resistance wire which yielded the height distribution and the frequency spectrum of the surface and optical glitter measurements which gave the upwind, downwind and crosswind slope distributions. The height

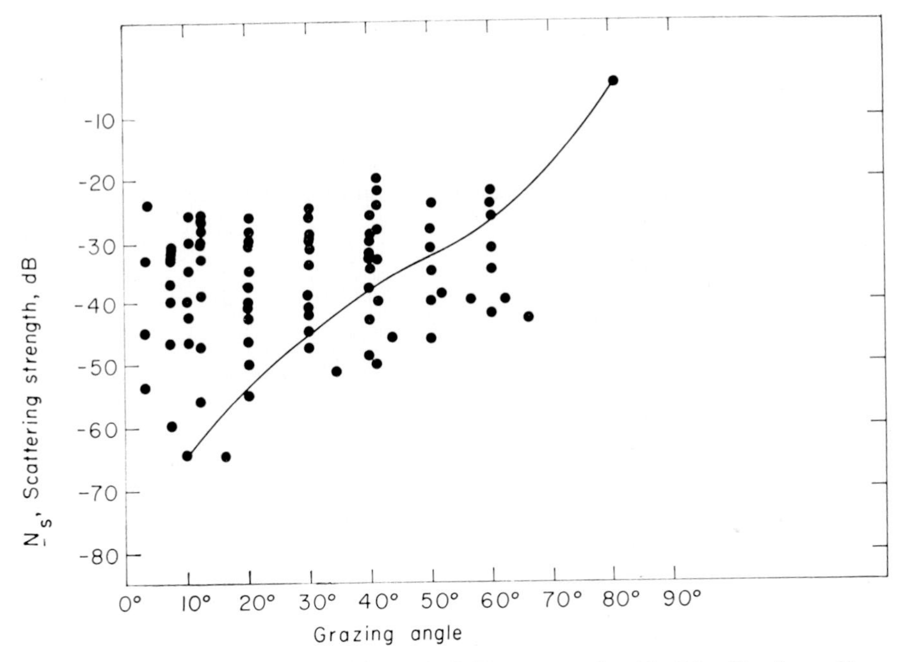

Fig. 3.5 Backscattering results[1, 15, 18, 20] compared with Marsh's theory[19]. (From Schulkin[14])

and slope distributions were nearly Gaussian, the latter closely resembled the data taken by Cox and Munk at sea. The data points for the sound intensities measured in the model experiment ($g \gg 1$) are shown in Fig. 3.6 with the curves predicted by the similar theories of Eckart and Beckmann. Again it is clear that the rough surface scattering theories are inadequate, although Eckart's prediction of the dependence of backscatter on the angle between the scattering plane and the wind direction is qualitatively correct. It is also interesting to observe that the levelling off of backscatter near grazing that has been observed at sea is not evident in the laboratory experiment where the experimental data are well above noise throughout 60 dB of monotonic change of level from normal to 80° incidence. The surface was rough ($g > 1$) for sound incidence from 0° to 75° for this experiment.

3.3.2 Two hypotheses to resolve the difficulties

How are we to explain the great divergence of theory from data points for the range of incidence angles from 30° to grazing? Two hypotheses offer promise.

(1) The complete ocean distribution of surface heights and the correlation function are not Gaussian. In particular, if one is interested in the scatter of sound wave lengths comparable to the capillary wave lengths ($\Lambda_{cap} \leqslant 2{\cdot}4$ cm) it is necessary to include these components explicitly in the description of the surface and to develop a theory for scattering from such a multi-surface. It would be expected on simplest physical arguments

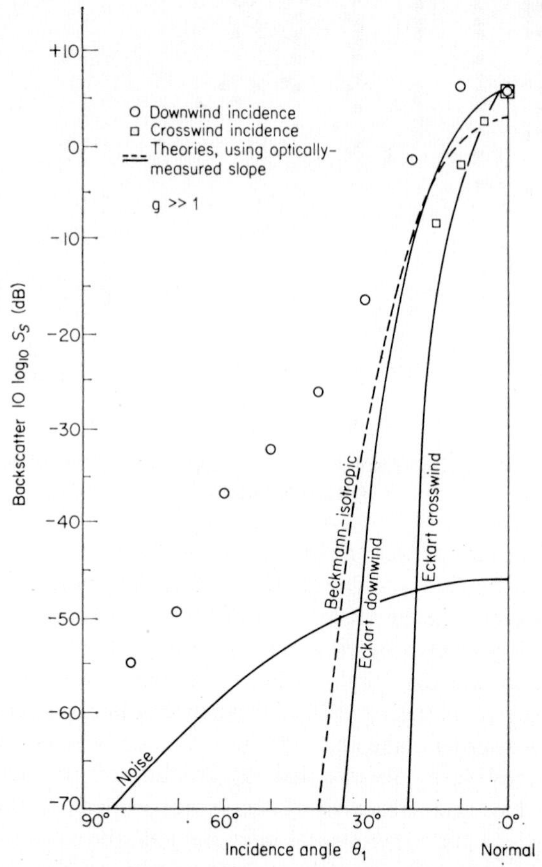

Fig. 3.6 Laboratory model data for backscattering from a rough surface[22]. $g = 21{\cdot}2 \cos^2 \theta$, $f = 450$ kHz

that if the sound wave length is comparable to or greater than the randomly oriented capillary wave lengths there will be supplementary scattering which will be nearly isotropic; such scattering will assume particular importance at angles nearer grazing where there is relatively little backscattering from the surface wave system. There have been several analytical attacks on the problem. An early study of a surface of small sinusoids superimposed on large sinusoids[23] showed that backscattering at near grazing angles was much greater than for a simple sinusoid. More recent results, for a statistically rough surface with small irregularities superimposed on large irregularities[24, 25] show that the backscattering will depend on the correlation between the two wave systems, and that scattering may not be simply additive. The predictions, based as they are on unknown correlation coefficients and unknown shadowing effects, are still unverified. The assumption of a Gaussian correlation is surely wrong[44].

(2) The second hypothesis[16, 25] assumes that there are enough scatterers of large scattering cross-section near the surface of the sea to give omnidirectional volume backscatter that is unresolvable from the surface backscatter; this will be the case particularly when the sound frequencies used are in the range of the resonant frequencies of existent bubbles due to rough seas or biological activity.

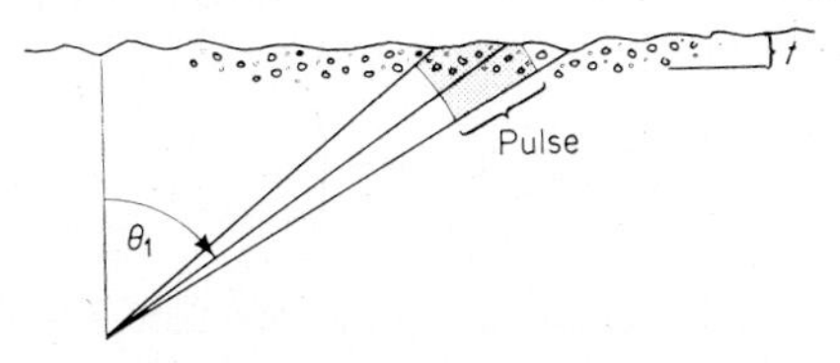

Fig. 3.7 Geometry of surface backscatter experiment in the presence of near-surface bubbles

If a surface backscattering experiment is performed at non-normal incidence it is clear (Fig. 3.7) that there will be a time at which backscatter is obtained simultaneously from the surface (left side of the beam) and from scatterers near the surface (from the right side of the beam which will not have yet reached the surface). The amount of this interference will be a function of pulse duration, beam width, and angle of incidence as well as the type of scatterer. As the angle of incidence is changed to become more nearly grazing the backscatter from the surface becomes even weaker until the volume backscatter from near-surface bubbles, which is omnidirectional, can dominate the total backscatter. The analytical statement for this effect can be derived by slightly modifying a fundamental development[1] for scattering from bubbles.

We assume a virtually plane wave of amplitude $p_1 = B/r_1$ entering a homogeneous, bubbly medium of depth t; the angle of incidence is θ_1. In the path length $dr = dz/\cos\theta_1$ the wave will attenuate due to resonant absorption and scattering by bubbles (the cross-sections for resonant bubbles are of the order 10^3 greater than those of nonresonant bubbles) so that

$$|p(z)|^2 = \frac{B^2}{r_1^2} \exp\left(-\frac{\sigma_e}{\cos\theta_1}\int_{z_0}^{z} n\,\mathrm{d}z\right) \tag{3.25}$$

where σ_e is the cross-section for bubble extinction (absorption plus scattering) at resonance and n is the number of resonant bubbles per unit volume at depth z. It is assumed that the bubble-layer thickness t is less than the pulse length and much less than r_1. Also, the acoustical signals scattered by each bubble are assumed to be incoherent so that the power scattered from a volume element $\mathrm{d}V$ is proportional to $n\sigma_s\,\mathrm{d}V$. The parameters are σ_s, the bubble-scattering cross-section at resonance, $\mathrm{d}V = r_1^2\,\mathrm{d}r\,\mathrm{d}\Omega$, the scattering volume, and $\mathrm{d}\Omega$, the solid angle subtended by the scattering volume. The scattered signal is attenuated as it backscatters through the bubble layer. The resulting incremental squared pressure backscattered within the solid angle, $\mathrm{d}\Omega$, is obtained by integrating over the total thickness of the bubble layer

$$\mathrm{d}|p|^2 = \frac{B^2\sigma_s}{8\pi r_1^2\sigma_e}\,\mathrm{d}\Omega\left\{1-\exp\left(-\frac{2\sigma_e}{\cos\theta_1}\int_0^t n\,\mathrm{d}z\right)\right\} \tag{3.26}$$

When the bubble layer is near a reflecting interface, (a) the reflector causes the bubbles to be ensonified from the top as well as from the bottom and (b) the forward-scattered sound is now reflected down so that it, too, reaches the hydrophone. As a first estimate we assume that these two effects increase $\mathrm{d}|p|^2$ by a factor of four. For very small numbers of bubbles per unit volume, and $\theta < 89°$, the exponent is very small and the approximation for the exponential yields

$$\mathrm{d}|p|^2 = \frac{B^2}{r_1^2}\,\frac{\sigma_s\,\mathrm{d}\Omega}{\pi\cos\theta_1}\int_0^t n\,\mathrm{d}z \tag{3.27}$$

Eqn. 3.27 must now be recast to reveal the surface-scattering factor due to bubbles, S_b. This factor is implied by any experiment in which it is not possible to resolve surface scattering in the presence of near-surface volume scattering. We calculate

$$S_b = \frac{\mathrm{d}|p|^2/|p_1|^2}{\mathrm{d}A/r_1^2}$$

using $\mathrm{d}\Omega = (\mathrm{d}A\cos\theta_1)/r_1^2$, and obtain

$$S_b = \frac{\sigma_s}{\pi}\int_0^t n\,\mathrm{d}z = \sigma_s N_s/\pi \tag{3.28}$$

where

$$N_s \equiv \int_0^t n \, dz \tag{3.29}$$

N_s is the number of bubbles per unit base area of the bubble column. The total surface scattering factor is assumed to be the sum

$$S = S_s + S_b. \tag{3.30}$$

Since S_b is independent of incidence angle, it places a lower limit on the values of surface-scattering factor, S, that may be obtained by an experiment in which near-surface bubbles exist.

Fig. 3.4 shows that a resonant bubble column $N_s = 10^{-2}$ cm^{-2} would explain Urick's backscatter data for incidence greater than 30°.

The factor of four introduced in Eqn. 3.27 has recently been considered more carefully by the author. If the bubble layer thickness is greater than the pulse length and if the width of the beam is now considered, the scattered intensity is independent of angle from normal incidence to about 70° but then droops significantly as grazing incidence is approached. This deviation from isotropic scatter (due to the nearby surface) agrees qualitatively with experiment and is a function of pulse length and source distance from the surface.

It is a pity that discussion of the bubble hypothesis must be concluded with the recognition that there are not enough measurements to make quantitative statements about bubbles at sea as a function of all the parameters that we would be interested in: for example, sea state, biological activity, depth below surface. We have only what the political commentators would call 'scattered and incomplete returns'.

Specifically the data are:

(1) Bubbles have been caught in a box at a depth of 10 cm in breaking waves near shore and the distribution found in 50 micron radius bands. The majority of the bubbles were less than 100 microns with bubble densities as high as 100 cm^{-3} for 30 micron radius and 10^{-1} cm^{-3} for 200 micron radius bubbles. Snowflakes were found to produce bubbles most prominently at 20 micron radius. Raindrops also produced bubbles[26].

(2) The surface of a laboratory water tank was agitated by fans to cause whitecaps; bubbles were caught in a tube and population was found to be centred at bubbles of radius 60 microns, i.e., resonant frequency 53 kHz[27].

(3) Bubbles have been photographed *in situ* just below water surface and at 10 ft depth 50 ft from shore[28].

(4) 'Bubbles' of radius 20 to 150 microns have been detected acoustically

in well-mixed shallow water of San Diego Bay in fairly uniform distributions with peaks of approximately 2 bubbles/litre at 30 kHz $\pm$ 10 per cent and 50 kHz $\pm$ 10 per cent and with lesser populations at depths down to 60 ft. (Calculations based on assumption that 70 per cent of cross-sections for a mixture of bubbles is caused by bubbles within 10 per cent of the resonant frequency[29, 45].)

(5) A physonectid siphonophore, possessing a gas float of radius of the order of 0·5 mm (resonant frequency approximately 10 kHz at sea level), has been shown to be capable of rapidly sinking by emission of gas bubbles and rising by efficient regeneration of its float bubble. Since siphonophores make up a large part of the plankton population in warm oceans they, and their emissions, have been postulated as the most common biological bubble scatterers in these waters[30, 31].

A factor which has not yet been mentioned is the question of bubble persistence. The full picture of free bubble populations at sea requires a knowledge of bubble production, dissipation and kinematics. Bubbles are lost by rising to the surface and by diffusing their gas into the surrounding water; the first effect removes large bubbles more effectively (greater than 60 micron radius), the second effect eliminates the smaller bubbles. In fact the particular prominence of 60 micron bubbles, resonant at approximately 55 kHz, is attributable to a minimum for these two co-operating effects.

Laboratory studies of rise time of bubbles in still water yield agreement with theory for terminal velocity under the influence of buoyant force and Stokes' drag. These studies show that a 60 micron bubble rises with a speed of approximately 1 cm s^{-1}. However, turbulent velocities as great as 5 cm s^{-1} (0·1 knot) are quite common at sea and a 60 micron bubble, entrained by such a turbulent motion, would *not* move towards the surface with a speed of 1 cm s^{-1} but would persist for much longer than one concludes from laboratory measurements.

3.3.3 *Other approaches*

It was Eckart[13] who first showed that, for relatively smooth surfaces ($g < 1$), the local phase shift at the surface should be rewritten as

$$e^{iv_z\zeta} \simeq 1 + iv_z\zeta$$

thereby immediately separating the coherent from the incoherent scattering component. Very, very briefly, the incoherent scattered intensity, which is all that remains in the case of backscatter, can then be stated in terms of the two-dimensional Fourier transform of the correlation function. This transform is the spectrum of the surface in wave number space. Finally, we assume that Bragg-like diffraction scattering will take place from the particular surface spectral component selected by the grating equation $2\Lambda \sin \theta_1 = \lambda$. An extension of this description of scattering leads to a prediction that there will be a frequency spreading of the scattered sound due to Doppler shift caused by movement of the appropriate 'diffraction grating'. There is conflicting experimental verification[32] and contradiction[33] of this prediction[46, 47]. The observed frequency broadening in the spectrum may be due to non-linear characteristics of the sea surface.

For $g < 1$, Eckart's solution for backscatter from an isotropic surface reduces to

$$S_s = \frac{2}{\pi} k^4 \cos^4 \theta_1 G'(2k \sin \theta_1)$$

where $G'(2k \sin \theta_1)$ is the energy density of the appropriate component of the two-dimensional surface wave system, given by the diffraction equation. The accuracy of this as well as other predictions has been tested in a laboratory experiment[22] in which simultaneous measurements were made of the surface spectrum at a point $G'(\Omega)$, the height distribution and the consequential sound scattering (Fig. 3.8). Conversion from the surface spectrum measured at a point to the two-dimensional surface spectrum in K space was made by assuming an isotropic system defined by

$$G'(K)\, K\, \mathrm{d}K = G(\Omega)\, \mathrm{d}\Omega$$

Another solution to the problem of scattering, again based on the spectral description of the sea surface, but assuming that the scattered field can be represented as a superposition of plane waves, has been derived[34] and applied to the cases of forward scattering in an isothermal surface channel[35] and non-specular scattering[19]. It is not possible to review the derivation at this time; however, Fig. 3.5 compares the previously-mentioned experimental data for backscatter at sea[1, 15, 18, 20] obtained at frequencies from 400 Hz to 60 kHz, with Marsh's theory. A Neumann–Pierson spectrum was assumed to describe the sea. Two difficulties are apparent: the fact that the experimental values are higher than Marsh's predictions at angles nearer grazing may be explained by the possible presence of volume scatterers; the magnitude and curvature of the theoretical curve near normal incidence is more disturbing. Marsh's assumption of plane wave superposition leading to position-independent amplitudes has been criticized[36] for the case of sinusoidal surfaces.

Fig. 3.8 shows the reasonably good agreement obtained for the theories tested (Eckart, Beckmann, Marsh) for the relatively smooth, wind-agitated, known water surface in the laboratory study.

An empirical model that appears to yield the correct forms of the

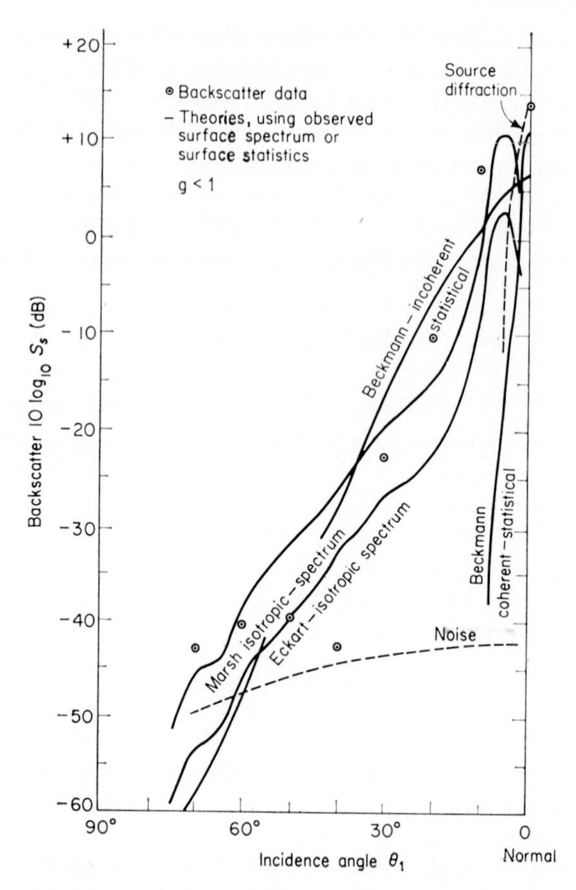

Fig. 3.8 Backscattering from a relatively smooth surface. Data points for laboratory experiment are compared with predictions from smooth surface theories of Eckart, Marsh and Beckmann[22]. $g=0{\cdot}5\cos^2\theta$, $f=70$ kHz

backscatter dependence on angle of incidence has been developed[37]. The scheme is to sum the effect of a large number of line scatterers of different sizes, slopes and directions, setting up the scattered intensity in the form

$$I(\theta)=\sum_{m=-\infty}^{\infty}\sum_{L=0}^{\infty} D(m)\,f(L)\,B(\theta, L)\,d(L) \tag{3.31}$$

where $D(m)$ specifies a Gaussian distribution of slopes of facets of the water surface; (L) is the 'target strength' taken as simply equal to L the length of a facet; $B(\theta, L)$ is the beam pattern of a line reflector (i.e., the diffraction pattern of a slit); and $d(L)$, the distribution of facet lengths, is specified as Rayleigh type. Fig. 3.9 taken from the paper shows that a computer program based on these assumptions leads to curves that resemble the experimental results. In reading the figure it should be noted that increasing the mean facet length at constant r.m.s. slope corresponds to increasing the

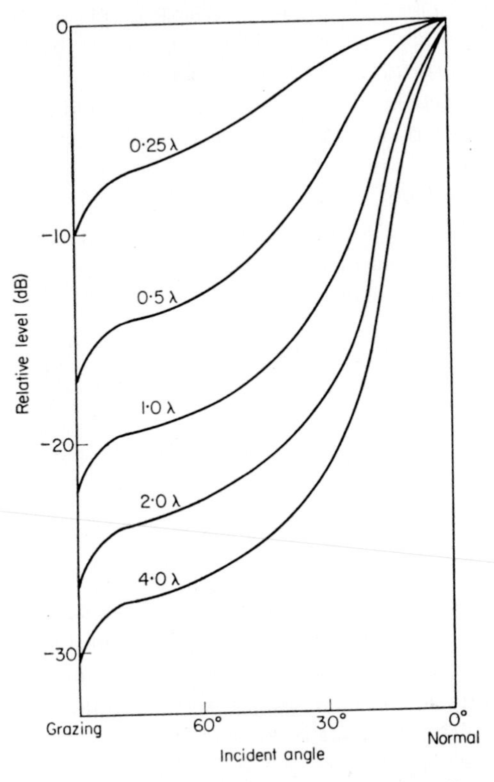

Fig. 3.9 Computer calculated backscatter curves for one-dimensional facets of r.m.s. slope 0·1 and of various lengths from 0·25 λ to 4·0 λ plotted as a function of grazing angle of incident sound[37]

correlation distance and the r.m.s. height; therefore the larger facet lengths represent greater roughness, g, in our nomenclature. One can conclude from this exercise that secondary mechanisms (such as volume scattering) *may* not be necessary to explain some surface reverberation experiments. In fact, similar experimental curves are found for radar backscatter, light backscatter from a mat surface and sound backscatter from an ocean bottom so that it is clear that a facet explanation which takes account of small surfaces will suffice for a large body of backscatter experiments in physics.

3.4 Specular scattering

3.4.1 The mean square scattered pressure

The acoustic intensity that reaches distant points is strongly dependent on the forward scattering experienced by any part of the sound wave which strikes the rough surface. The most important of the forward scattering directions is the specular direction which a ray would follow if the surface had been a perfect plane reflector. We can extract the case of 'specular scatter' for a piston source by setting $\theta_2=\theta_1$ and $\theta_3=0$ in Eqn. 3.21. In the specular direction $v_x=v_y=0$ and $F=1$.

Furthermore if $(u_x^2+u_y^2)\,L^2/16 \ll g$, which is commonly true for man-made sounds at sea, the predicted mean specular scattered intensity (relative to plane surface reflection) can be written as

$$\langle RR^* \rangle = e^{-g}\left(1+\frac{\pi^3 L^2 u_x u_y}{64}\sum_{m=1}^{\infty}\frac{g^m}{m!m}\right)$$

$$= e^{-g}+\frac{\pi^3 L^2 u_x u_y}{64}\,S(g) \tag{3.32}$$

where

$$S(g)=e^{-g}\sum_{m=1}^{\infty}\frac{g^m}{m!m} \tag{3.33}$$

The first term, which decreases rapidly with increasing surface roughness, represents the coherent component of the relative mean scattered intensity. The second, diffuse scattering term, which is comprised of randomly phased contributions, correspondingly grows with increasing roughness to an asymptote which is practically realized at $g>10$.

A graph of $S(g)$ is presented in Fig. 3.10 from which it is clear that we may use the approximations

$$\underset{g\ll 1}{S(g)\to g} \qquad \underset{g\gg 1}{S(g)\to g^{-1}}$$

so that the limiting, specularly-scattered, relative intensities for very smooth and very rough surfaces are

$$\langle RR^* \rangle = e^{-g} \qquad g \leqslant 0\cdot 1 \tag{3.34}$$

$$\langle RR^* \rangle = \frac{\pi^3 A'}{512\Sigma^2 r_1^2 \cos^2\theta_1} \qquad g \geqslant 10 \tag{3.35}$$

where $A'=4BD$ is the area of the piston source.

Note: The comparable form for the diffuse term in the case of *uniform* ensonification over a *surface area A* is readily derivable[11] as

$$\langle RR^* \rangle = \lambda^2/(8\pi A \Sigma^2 \cos^2 \theta_1)$$

Eqn. 3.34 and 3.35 state that for a relatively smooth homogeneous, isotropic Gaussian surface the mean specularly scattered intensity (relative to that from a plane surface) depends only on the r.m.s. height σ (through g); for a relatively rough surface the intensity for a given sound source, position and angle of incidence depends only on the r.m.s. slope of the surface, Σ. If these equations properly describe the effect of the sea, it should

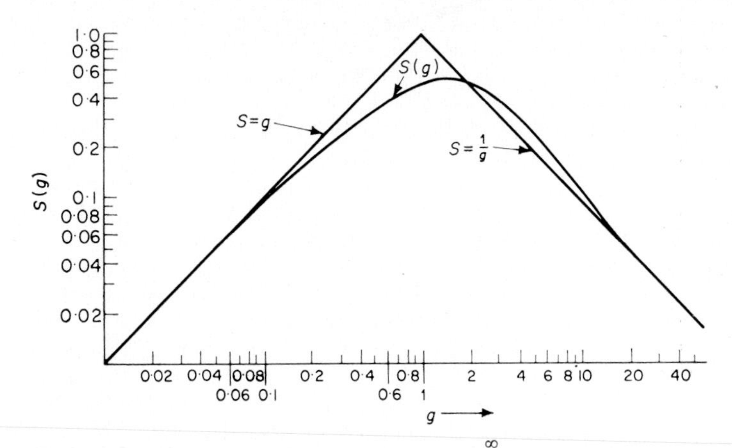

Fig. 3.10 Graph of scattering function $S(g) = e^{-g} \sum_{m=1}^{\infty} g^m/mm!$ for the incoherent component of scattered intensity at specular scatter

be possible to determine the r.m.s. wave height σ from a specular scattering experiment at low frequencies and to find the mean square slope of the water surface by the results of a similar experiment at high frequencies. To my knowledge, no experiment to test this theory has as yet been conducted at sea. However a careful study[38] using a pressure release surface constructed of one-dimensional Gaussian distributed heights of cork floating on the water surface has shown excellent agreement with theory particularly for $g \lesssim 1$ for angles of incidence 0° to 60°. We have recently made simultaneous glitter measurements of wind-agitated, near-Gaussian, water slopes and specularly scattered 20–200 kHz sound at normal incidence[21, 22] in a large anechoic tank and have found excellent agreement between theory and experiment from $g = 0{\cdot}01$ up through the asymptote at large $g (= 60)$. Fig. 3.11 shows the contributing coherent and incoherent components. The slope measured by optical glitter was only

17 per cent greater than the value obtained from high frequency specular scatter. In fact the model experiment shows agreement with theory even in the 'forbidden' region $L \simeq \lambda$ which the Kirchhoff assumption does not claim to cover (recall $L \gg \lambda$ in the theory). On the evidence of these model experiments as well as the work of Proud[38] it can be stated that the theory

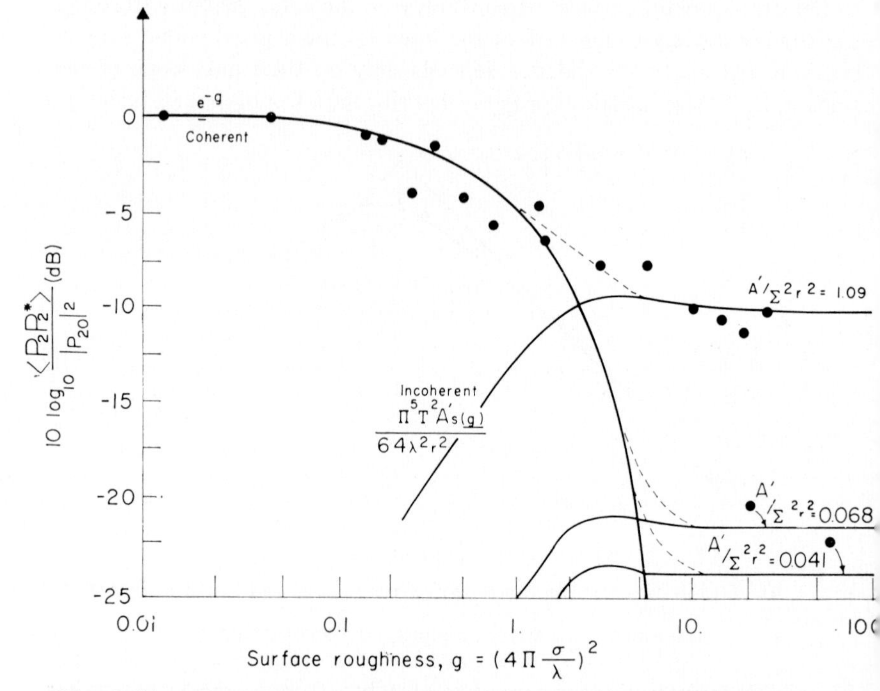

Fig. 3.11 Mean square specularly-scattered pressure at normal incidence, relative to mirror intensity, for different surfaces. The coherent component follows the law e^{-g} but the incoherent term has an asymptote, for large g, that is a function of the geometry. Data for Fig. 3.11 were obtained at frequencies 20–450 kHz, r.m.s. heights 0·059 to 0·31 cm, and mean square slopes 0·011 to 0·033

outlined here provides reasonably accurate predictions of specularly scattered r.m.s. intensities, from a knowledge of r.m.s. height and r.m.s. slope for a near-Gaussian water surface, provided that the incidence is not close to grazing.

With this verification, we now have the tools to return to the rough surface Lloyd's Mirror problem as discussed in the Introduction. It is clear that the time-average solution must be calculated in two parts; the direct radiation from the source (Eqn. 3.2) will interfere with the coherent component of the specularly scattered sound; superimposed on this partial interference pattern there will be an incoherent component which will add to both the peaks and the troughs of that pattern. Both components of the scattered sound will be functions of the angle of incidence.

3.4.2 *The distribution of values of scattered pressure*

The scattered field that we have been discussing is the mean square of the relative acoustic pressure. As the surface fluctuates, so does the instantaneous scattered sound pressure. An experiment at sea is confounded by temperature fluctuations and moving volume inhomogeneities so that the variations due to the surface cannot be extracted. These complications are avoided in the laboratory. Fig. 3.12 shows the scatter of acoustic data

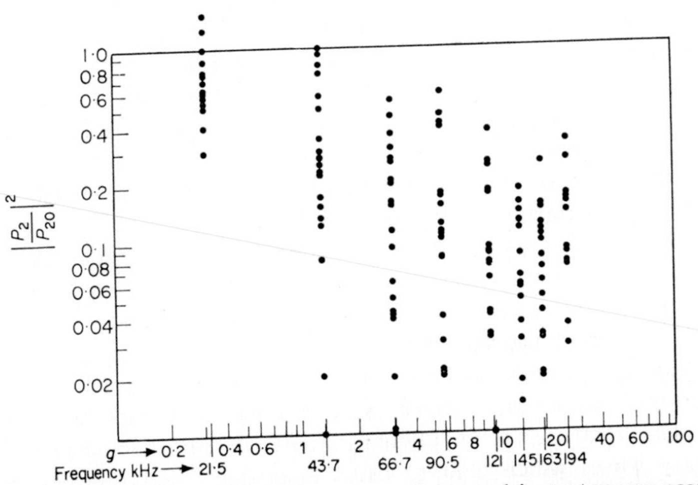

Fig. 3.12 Laboratory experimental values of squared instantaneous scattered pressure relative to smooth surface reflected pressure for normal incidence as a function of sound frequency and the roughness parameter[21]

due to surface fluctuations alone, observed in the previously-mentioned anechoic tank experiment. The statistical distribution of such an experiment has been studied, particularly by Beckmann[11] who shows that the specularly scattered pressure distribution will range from near Gaussian with large mean value for relatively smooth surfaces ($g \ll 1$) to Rayleigh (Eqn. 3.36) for rough surfaces ($g \gg 1$). The scattered pressures in the non-specular directions are predicted to be Rayleigh distributed,

$$w(x) = \frac{x}{S^2} \exp\left[-\frac{x^2}{2S^2}\right] \tag{3.36}$$

The Gaussian distribution is caused by the vector addition of a strong constant vector, $\langle R \rangle$, and a small variable vector with real and imaginary components which are Gaussian distributed but of possibly unequal variances. The Rayleigh (rough surface) distribution, on the other hand, occurs when the constant vector becomes negligible and there remains only the distribution due to small, equal vector components, with equally probable phases. Beckmann shows further that 'of all the possible distributions of the amplitude of a field scattered by a symmetrically distributed rough surface the Rayleigh distribution is the one with the greatest variance'.

These predictions have been verified in the laboratory[22]. As the surface roughness, g, increased there was a shift in probability distribution from near-Gaussian with large mean value $\langle |p/p_{20}|^2 \rangle$ at $g=0 \cdot 12$ to near-Rayleigh with smaller mean value at $g=1 \cdot 5$ (see Fig. 3.13). For the near-Gaussian distribution the variance was a much smaller fraction of the mean square value than for the near-Rayleigh distributions. The theoretical value of the variance for a true Rayleigh distribution is 21·2 per cent of the mean square value; the experimental variances for the cases of rough surface scattering ($g > 3$) were an average of approximately 20 per cent of the experimental mean square values.

To temper the satisfaction and confidence that comes with the above demonstrations of the effectiveness of the Helmholtz theory for specular scatter from a Gaussian surface it should be recalled that:

(a) The importance (or the insignificance) of shading and multiple reflection for specular scattering at near-grazing incidence has not been proven for the Gaussian surface. An attack has been made by Wagner[39].

(b) The 'total' sea is undoubtedly non-Gaussian and the presence of capillary waves which tend to cause peakedness and skewness to the distribution may be important for forward scatter when the sound wave length is comparable to the capillary wave length (f acoustic $\geqslant$ 30 kHz). At wind speeds greater than about 5 m s^{-1} the r.m.s. slope of the surface (as measured optically) is strongly dependent on the contributions from capillary waves[40] so that the value of Σ to be used in theory is no longer obvious. Furthermore, long wave length swell propagated from distant storms will have the effect of tilting the plane for reference for the surface heights and thereby will influence both the distribution and the r.m.s. value of the specularly scattered pressure for all except perhaps infrasonic frequencies.

(c) The study of scattering by an anisotropic sea, which in fact is the typical sea, has hardly been mentioned in the scientific literature, let alone solved and compared with experiment.

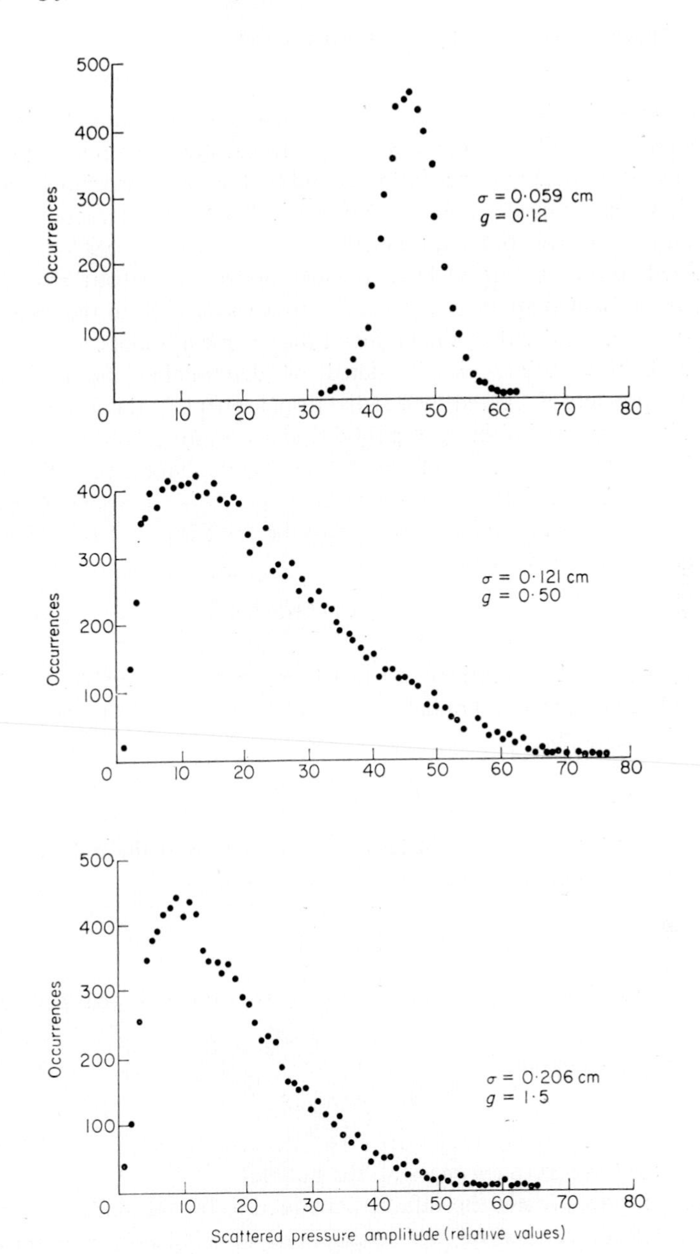

Fig. 3.13 Probability densities for normal incidence backscatter for three different surfaces[22]. $f = 70$ kHz

3.5 Appendix: Scattering by resonant bubbles

3.5.1 *Resonant frequency*

From the simplest acoustical point of view, a small gas bubble in water is comprised of stiffness (measured by the radial stress required to produce a change in radius) and inertia (measured by the mass associated with the bubble when it pulsates about its position). These two characteristics are, of course, the essential requirements of an oscillatory system, and the localized constants that we have assumed above are realistic because the system resonant frequency corresponds to a wavelength in the water that is very much greater than the radius of the spherical bubble.

The bubble stiffness is calculated by determining the incremental internal pressure P_i caused by a radial displacement ξ; this is obtained by assuming that the bubble gas pulsates *adiabatically* according to $PV^{\gamma}=$ constant where γ is the ratio of specific heats for the bubble gas. Differentiation yields the relation $P_i=-3\gamma P_0\xi/R_0$ where P_0 is the ambient pressure and R_0 is the mean bubble radius. Over the full surface of the sphere the force assumes the Hookesian relation

$$F_{\text{stiffness}}=-\underbrace{(12\pi\gamma P_0 R_0)}_{s}\,\xi$$

where the factor in parentheses is the 'stiffness' (s) of the bubble.

A pulsating bubble will radiate as a simple source producing the scattered field of a monopole

$$P_s=\frac{B}{r}\,e^{j(\omega t-kr)}$$

To evaluate the constant B in terms of the motion at the bubble surface we set up the momentum equation for radial motion

$$\rho_0\ddot{\xi}=\left[-\frac{\partial P_s}{\partial r}\right]_{r=R_0}$$

which yields $B=-\rho_0 R_0^2|\ddot{\xi}|$. Finally the inertial force is identified as

$$F_m=-P_s 4\pi R_0^2$$

from which we obtain

$$F_m=\underbrace{4\pi R_0^3\rho_0}_{m}\ddot{\xi}$$

where m is the equivalent mass of 'the bubble'.

We can now present the differential equation for the unforced motion of the bubble surface on the assumptions of adiabatic conditions, no surface tension, no dissipative forces

$$4\pi R_0^3\rho_0\ddot{\xi}+12\pi\gamma P_0 R_0\xi=0$$

From this we obtain the adiabatic resonant frequency for a pulsating bubble

$$f_A = \frac{1}{2\pi}\sqrt{\frac{s}{m}} = \frac{1}{2\pi R_0}\sqrt{\frac{3\gamma P_0}{\rho_0}}$$

For an air bubble at sea level this formula reduces to

$$f_A(\text{Hz}) = \frac{320}{R_0(\text{cm})}$$

The effect of surface tension comes in significantly only if the radius is less than 10 microns. The assumption of adiabatic gas behaviour is no longer valid for minute bubbles in which there is intimate contact between the gas and the water so that expansions and compressions become isothermal. The cross-over in behaviour occurs for bubbles near 40 microns. Otherwise stated, the *adiabatic* resonant frequency is in error by less than 5 per cent for bubbles of radius greater than 60 microns, and the *isothermal* resonant frequency is very nearly correct for bubbles of radius less than 10 microns. The details of these corrections may be found in the literature[41].

3.5.2 *Scattering and absorption cross-sections*

Now let us consider what will happen when a plane wave is incident at the bubble. We assume that the incident sound is of wave length very much greater than the bubble radius, R_0, that is $kR_0 \ll 1$. The incident acoustic pressure will then have virtually the same value at all points over the surface of the bubble; we write this forcing pressure

$$P_f = Ae^{j\omega t}$$

and the corresponding incident intensity

$$I_f = \frac{A^2}{2\rho_0 c}$$

where $\rho_0 c$ is the acoustic impedance of the water medium. The bubble will be forced to pulsate at the incident frequency and will re-radiate (scatter) according to

$$P_s = \frac{B}{r} e^{j(\omega t - kr)}$$

where $\boldsymbol{B}$ must now be complex because of the possible phase shift relative to the incident pressure. The scattered intensity is

$$I_s = \frac{|B|^2}{2r^2 \rho_0 c}$$

The important concept of the total scattering cross-section, σ_s, is now introduced by defining

$$\sigma_S = \frac{\text{total scattered power}}{\text{incident intensity}}$$

For the bubble scattering cross-section we get

$$\sigma_S = 4\pi \frac{|B|^2}{A^2}$$

This ratio is evaluated by matching the internal bubble pressure and the radial bubble velocity to the corresponding values in terms of the scattered field at $r = R_0$. The complete solution requires a consideration of the thermal conductivity and shear viscosity at the bubble walls, a subject too involved for the present development. The details are reviewed by Devin[41]. The conclusion can be presented in the expected form for a damped, resonant system

$$\sigma_S = \frac{4\pi R_0^2}{[(\omega_0/\omega)^2 - 1]^2 + \delta^2}$$

where ω_0 = bubble resonant frequency and $\delta = \delta_r + \delta_v + \delta_t$ is the bubble damping constant due to re-radiation, shear viscosity and thermal conductivity.

The preceding equation leads to an impressive conclusion: *at resonance*, the scattering cross-section is limited only by the value of the damping

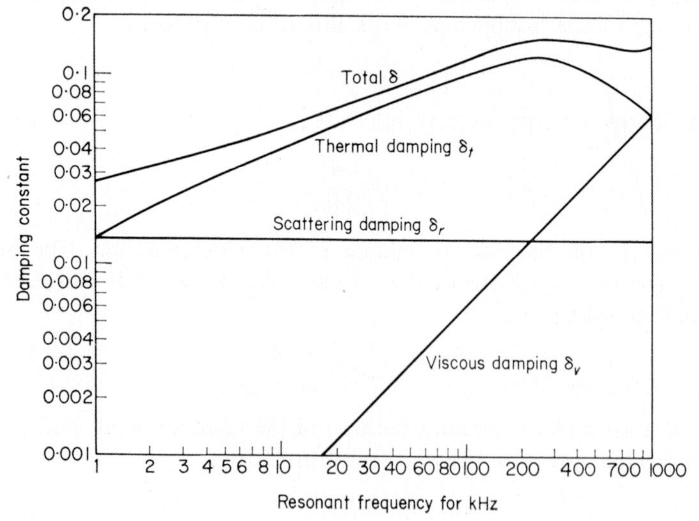

Fig. 3.14 Theoretical damping constants for resonant air bubbles in water[41]

constant (see Fig. 3.14); since this constant has fractional values, σ_s can be vastly greater than the geometrical cross-section of the bubble, πR_0^2. For example the ratio of the acoustical scattering cross-section to geometrical cross-section at resonance is 400 for a 60 kHz bubble and 5000 for a 1 kHz bubble. A simple physical explanation of this effect is that the resonant bubble, being a very low impedance region, will distort the equiphase surfaces of the incoming plane wave over a region very much larger than the geometrical size of the bubble itself.

Another useful concept is the absorption cross-section, defined as

$$\sigma_a = \frac{\text{total power absorbed}}{\text{incident intensity}}$$

Since it can be shown that

$$\frac{\sigma_a}{\sigma_s} = \frac{\delta_v + \delta_t}{\delta_r}$$

the relative values of the damping constants plotted in Fig. 3.14 indicate that above 1 kHz, $\sigma_a > \sigma_s$.

Finally the extinction cross-section, σ_e is introduced by the definition

$$\sigma_e = \sigma_a + \sigma_s$$

It is seen that σ_e measures the power taken from the sound beam by both scattering and actual absorption (conversion to thermal energy).

References

1 *Principles of Underwater Sound* (C. Eckart, ed.). U.S. Office of Scientific Research and Development, National Defense Research Committee, Div. 6, Vol. 7. Distributed by National Research Council, Washington, D.C. (1951).

2 Cox, C. S., and W. H. Munk, 'Measurement of the Roughness of the Sea Surface from Photographs of the Sun's Glitter', *J. Optic. Soc. Am.*, **44**, 838–850 (1954).

3 Cox, C. S., and W. H. Munk, 'Statistics of the Sea Surface Derived from Sun Glitter', *J. Marine Res.*, **13**, 198–227 (1954).

4 Kinsman, B., *Wind Waves*, p. 345 (Prentice-Hall, Englewood Cliffs, N.J.) (1965).

5 MacKay, J. H., 'The Gaussian Nature of Ocean Waves, Project A-366', *Internal Technical Note*, No. 8, Eng. Expt. Station, Georgia Inst. of Technology, Atlanta, Georgia (1959).

6 *Ocean Wave Spectra*, Proceedings at a conference, National Academy of Sciences (Prentice-Hall, Englewood Cliffs, N.J.) (1963).

7 Pierson, W. J., Jr. and L. Moskowitz, 'A Proposed Spectral Form for Fully Developed Wind Seas Based on the Similarity Theory of S. A. Kitaigorodskii', *J. Geophys. Res.*, **69**, 5181–5190 (1964).

8 Pierson, W. J., Jr., G. Neumann, and R. W. James, 'Practical Methods for Observing and Forecasting Ocean Waves by Means of Wave Spectra and Statistics', *Publ. U.S. Navy Hydr. Off.* No. 603 (1955).
9 Darbyshire, J., 'An Investigation of Storm Waves in the North Atlantic Ocean', *Proc. R. Soc. A*, **230**, 560–569 (1955).
10 Phillips, O. M., *The Dynamics of the Upper Ocean* (Cambridge U.P., Cambridge, England) (1966).
11 Beckmann, P., and A. Spizzichino, *The Scattering of Electromagnetic Waves from Rough Surfaces* (Pergamon, Oxford and Macmillan, New York) (1963).
12 Isakovitch, M. A., 'The Scattering of Waves from a Statistically Rough Surface', *J.E.T.P.*, **23**, 305–314 (1952).
13 Eckart, C., 'The Scattering of Sound from the Sea Surface', *J. Acoust. Soc. Am.*, **25**, 566–570 (1953).
14 Schulkin, M., and R. Shaffer, 'Backscattering of Sound from the Sea Surface', *J. Acoust. Soc. Am.*, **36**, 1699–1703 (1964).
15 Urick, R. J., and R. M. Hoover, 'Backscattering of Sound from the Sea Surface: Its Measurement, Causes and Applications to the Prediction of Reverberation Levels', *J. Acoust. Soc. Am.*, **28**, 1038–1042 (1956).
16 Clay, C. S., and H. Medwin, 'High Frequency Acoustical Reverberation from a Rough Sea Surface', *J. Acoust. Soc. Am.*, **36**, 2131–2134 (1964).
17 Chapman, R. P., and J. H. Harris, 'Surface Backscattering Strengths Measured with Explosive Sound Sources', *J. Acoust. Soc. Am.*, **28**, 1592–1597 (1962).
18 Garrison, G. R., S. R. Murphy and D. S. Potter, 'Measurements of the Backscattering of Underwater Sound from the Sea Surface', *J. Acoust. Soc. Am.*, **32**, 104–111 (1960).
19 Marsh, H. W., 'Nonspecular Scattering of Underwater Sound by the Sea Surface', Lecture 11, *Underwater Acoustics* (W. M. Albers, ed.) (Plenum, New York), pp. 193–197 (1963).
20 Chapman, R. P., and H. D. Scott, 'Surface Backscattering Strengths Measured Over an Extended Range of Frequencies and Grazing Angles', *J. Acoust. Soc. Am.*, **36L**, 1735–1737 (1964).
21 Medwin, H., 'Specular Scattering of Underwater Sound from a Wind-driven Surface', *J. Acoust. Soc. Am.*, **41**, 1485–1495 (1967).
22 Medwin, H., E. C. Ball and J. A. Carlson, 'Backscattering of Underwater Sound from a Wind-driven Model Sea', *J. Acoust. Soc. Am.*, **42,** 1184 (1967)
23 La Casce, E. O., Jr., 'Note on Backscattering of Sound from the Sea Surface', *J. Acoust. Soc. Am.*, **30**, 578–580 (1958).
24 Kur'yanov, B. F., 'The Scattering of Sound at a Rough Surface with Two Kinds of Irregularity', *Sov. Phys. Acoust.*, **8**, 252–257 (1963).
25 Urick, R. J., 'The Processes of Sound Scattering at the Ocean Surface and Bottom', *J. Marine Res.*, **15**, 134–148 (1956).
26 Blanchard, D. C., and A. H. Woodcock, 'Bubble Formation and Modification in the Sea and its Meteorological Significance', *Tellus*, **9**, 145–157 (1957).
27 Glotov, V. P., P. A. Kolobaev and G. G. Neuimin, 'Investigation of the Scattering of Sound by Bubbles Generated by an Artificial Wind in the

Sea Water and the Statistical Distribution of Bubble Sizes', *Sov. Phys. Acoust.*, **7**, 341–345 (1962).

28 Barnhouse, P. D., M. J. Stoffel and R. E. Zimdar, 'Instrumentation to Determine the Presence and Acoustic Effect of Microbubbles near the Sea Surface', *M.S. Thesis*, Naval Postgraduate School, Monterey, California (1964).

29 Buxcey, S., J. H. McNeil and R. H. Marks, Jr., 'Acoustic Detection of Microbubbles and Particular Matter near the Sea Surface, *M.S. Thesis*, Naval Postgraduate School, Monterey, California (1965).

30 Barham, E. G., 'Deep Scattering Layer Migration and Composition: Observations from a Diving Saucer', *Science*, **151**, 1399 (1965).

31 Pickwell, G. V., E. G. Barham, and J. W. Wilton, 'Carbon Monoxide Production by a Bathypelagic Siphonophore', *Science*, **144**, 860 (1964).

32 Lieberman, L. N., 'Analysis of Rough Surfaces by Scattering', *J. Acoust. Soc. Am.*, **35**, 932 (1963).

33 Mellen, R. H., 'Doppler Shift of Sonar Backscatter from the Sea Surface', *J. Acoust. Soc. Am.*, **36**, 1395–1396 (1964).

34 Marsh, H. W., 'Exact Solution of Wave Scattering by Irregular Surfaces', *J. Acoust. Soc. Am.*, **33**, 330–333 (1961).

35 Marsh, H. W., M. Schulkin and S. G. Kneale, 'Scattering of Underwater Sound by the Sea Surface', *J. Acoust. Soc. Am.*, **33**, 334–340 (1961).

36 Murphy, S. R., and G. E. Lord, 'Scattering from a Sinusoidal Surface—a Direct Comparison of the Results of Marsh and Uretsky', *J. Acoust. Soc. Am.*, **36L**, 1598–1599 (1964).

37 Patterson, R. B., 'Model of a Rough Boundary as a Back-scattering of Wave Radiation', *J. Acoust. Soc. Am.*, **36**, 1150–1153 (1964).

38 Proud, J. M., Jr., R. T. Beyer and P. Tamarkin, 'Reflection of Sound from Randomly Rough Surfaces', *J. Appl. Phys.*, **31**, 543–553 (1960).

39 Wagner, R. J., 'Shadowing of Randomly Rough Surfaces', *J. Acoust. Soc. Am.*, **41**, 138–147 (1967).

40 Cox, C. S., 'Comments on Dr Phillips' Paper', *J. Marine Res.*, **16**, 199–225 (1958).

41 Devin, C., 'Survey of Thermal, Radiation and Viscous Damping of Pulsating Air Bubbles in Water', *J. Acoust. Soc. Am.*, **31**, 1654–1667 (1959).

42 Schooley, A. H., 'A Simple Optical Method for Measuring the Statistical Distribution of Water Surface Slopes', *J. Opt. Soc. Am.*, **44**, 37–40 (1954).

43 Burling, R. W., 'The Spectrum of Waves at Short Fetches', *D. Hydrogr. Z.*, **12**, 45–64 and 96–117 (1959).

44 Medwin, H. and C. S. Clay, 'Dependence of Spatial and Temporal Correlation of Forward Scattered Underwater Sound on the Surface Statistics: Part II—Experiment', *J. Acoust. Soc. Am.*, **47** (May 1970).

45 Medwin, H., '*In situ* Acoustic Measurements of Bubble Populations in Coastal Ocean Waters', *J. Geophys. Res.*, **73**, 599–611 (1970).

46 Parkins, B. E., 'Scattering from the Time-varying Surface of the Ocean', *J. Acoust. Soc. Am.*, **42**, 1262–1267 (1967).

47 Marsh, H. W. and R. H. Mellon, 'Boundary Scattering Effects in Underwater Sound Propagation', *Radio Science*, 339–346 (1966).

4

An Introduction to Acoustic Exploration

W. F. Hunter
Physics Department, Imperial College of Science and Technology, London
(Now at R.A.N. Research Laboratory, Sydney, Australia)

4.1 Introduction

Most people practising underwater acoustics do so in naval laboratories, in geological surveying, in oceanographic institutes or in universities. In the universities, a wide range of cavitation phenomena have been studied in all sorts of water and there is some industrial interest in applications of *ultrasonic* cavitation; but this is a small fraction of the total applications of underwater acoustics. Almost all the rest is practised in sea water; mainly as a tool for exploration of one kind or another. The main acoustic technique is echo analysis which is used by the geologists in locating and studying earth structures, by biologists in studying fish and by ships in detecting submarines. The bulk of professional underwater acousticians are working directly on naval defence problems in naval establishments, in industry and in some universities. The intention of this chapter is to give a brief introduction to acoustic exploration at sea.

Above 50 kHz, the absorption of sound in sea water increases as f^2 and the background thermal agitation noise intensity increases as f^2 so the useful range of propagation falls very rapidly with increase of frequency (f) above 50 kHz. Frequencies as high as 500 kHz are used occasionally at sea, but only where high resolution at very short range is required such as in some studies of fish. The highest frequencies are used in velocimeters, which work at 3–5 MHz over path lengths up to 20 cm. As far as ocean acoustics is concerned there is no such thing as a frequency of 10 MHz because the absorption would be more than 30 dB m^{-1}.

Practical long range exploration makes use of the frequency band 10 Hz to 20 kHz, so the remarks in this paper will apply mainly to this range which is usually called the audio frequency band.

4.1.1 Definitions

Acoustic exploration This concerns the use of acoustic techniques to answer questions such as the following.

(a) How deep is the water under a ship and what is the nature of the sea bed?

(b) How many fish are under the boat now; what is their depth and what sort of fish are they?

(c) Is there a submarine near the ship; if so where is it, which way is it going and how fast?

(d) Is the stationary object on the bottom, a rock or a mine?

(e) Is a torpedo running towards the boat, if so, from what direction?

As an example of using acoustic techniques; if a microphone picks up a loud bang, the answer to the last question was 'Yes'. And incidentally, this is a perfect example of the correct answer being given too late—a common problem in signal processing, as in life.

Target This is the specific object that the *observer* wishes to detect.

Exploration is split into active and passive detection. The principle of *passive detection* is that the *target* must radiate energy into a medium—either directly or scattered from some natural source like sunlight. The observer intercepts some of the radiated (or re-radiated) energy and must make his deductions about the target based on the received energy which has been modified both by propagation through the medium and by the characteristics of the observer's equipment. Good examples of passive detection are seismic deductions about earthquakes and normal visual observations in sunlight. Passive detection from scattered sound is seldom exploited in underwater acoustics, so in practice passive detection means listening to energy radiated by the target.

In *active detection* the *observer* radiates energy into a medium. This energy is modified by the medium and by the target and the observer makes his deductions about the target from the scattered energy which he intercepts. For instance a searchlight uses continuous active detection. The light is scattered back to an observer's eye by any targets in the beam. Targets out of the beam are not detected. Continuous energy radiation is not generally used for sonar operation because more value is obtained from intermittent insonification called *echo ranging* in which the observer sends out pulses of energy called *pings*, then listens for echoes. *Classification* is the identification of a specific target, based on all the available evidence.

4.2 Echo ranging

4.2.1 The sonar equations

Traditionally, echo ranging has been described by the *sonar equations* analogous to the *radar equation*. The popularity of these equations presumably stems from the fact that one can play games lumping together 'like' terms and hiding terms which you don't like, or better still allocating them to someone else to investigate. There are difficulties in using these equations but first consider the ideas behind them. By tradition, intensity is regarded as the parameter to juggle with and manipulations are carried out in decibels. Absolute quantities will be denoted in lower case and their logarithms in upper case, e.g., $S = 10 \log s$. If a signal of intensity S_A is sent out and an echo of intensity S_E is received, then for a target of unit strength

$$S_E = S_A - P_2$$

where P_2 is the two-way propagation loss. This is a definition of P_2. That is, the amount we get back (S_E) is the amount we sent out (S_A) less the amount lost on the way (P_2). Note that intensity lost does not necessarily mean energy dissipated. Energy which deviates by scattering and refraction is also lost to the detection system.

Some of the energy 'lost on the way' is lost at a single point in the propagation path, namely *at* the target. It is more convenient to account for this separately, for which we introduce the definition *target strength*. The target strength is simply the fraction of the intensity incident at the target which is scattered back in the direction of the observer. If T is the target strength in decibels, then

$$S_E = S_A - P_2 + T \tag{4.1}$$

Now, P_2 is reserved for the propagation loss, other than intensity changes *at the target*. We will investigate the meaning of 'at the target' later.

Traditional reasoning is that the minimum recognizable value of S_E will occur at some fraction of the noise (N) at the receiver, which gives us a *recognition differential* (R) such that

$$S_{E(min)} = N + R \tag{4.2}$$

Then any other echo level is given by

$$S_E = S_{E(min)} + S_X \tag{4.3}$$

where S_X is the excess echo level available. Then from Eqns. 4.1, 4.2 and 4.3 we get the *echo ranging equation* (*for active sonar*)

$$S_A - N - R - P_2 + T = S_X \tag{4.4}$$

By similar reasoning, one obtains the *sonar listening equations* (*for passive sonar*)

$$S_P - N - R - P_1 = S_X \tag{4.5}$$

where S_P is the radiated noise of the target and P_1 is the one way propagation loss.

There are many variants of these intensity equations, probably the most notable is the treatment by Urick[1] in which energy densities are equated instead of intensities.

4.2.2 *Figure of merit*

The figure of merit (F) for a sonar set gives the maximum allowable two-way propagation loss for a threshold detection; i.e., if $S_X = 0$ then

$$F = P_2 - T \tag{4.6}$$

which is a function of range, so that a set with a high figure of merit will have a larger allowable two-way propagation loss and thus a greater detection range.

From Eqns. 4.4 and 4.6

$$F = S - N - R \tag{4.7}$$

that is, F should increase as the source intensity increases and as noise and recognition differential decrease. The idea was that F could be measured from Eqn. 4.7 to rate the qualities of different designs of sonar sets and then from Eqn. 4.6 detection range as a function of propagation conditions could be determined. Sometimes Eqn. 4.7 is further factorized

$$N = N_1 - D + 10 \log b$$

where N_1 is the equivalent isotropic noise spectrum level (by convention spectrum level means the intensity in a 1 Hz band), D is the directivity index and b is the bandwidth of the receiver.

4.2.3 The argument for S_A, P_2 and T

The sonar Eqns. 4.4 and 4.5 have been 'derived' on a plausible argument, but it is necessary to look more closely into the matter to find the exact meanings of the terms. There are many ways of doing this and the meanings may differ as a result. We will pursue a representative argument based on uniform geometrical spreading and uniform absorption. Take an echo ranging situation with a source at A and a target at B separated by a distance r (Fig. 4.1). S_A is the active source intensity at distance r_A from A. S_B is the incident intensity at the target B and S_T is the reflected intensity at a distance r_B from the target. S_E is the echo level at A.

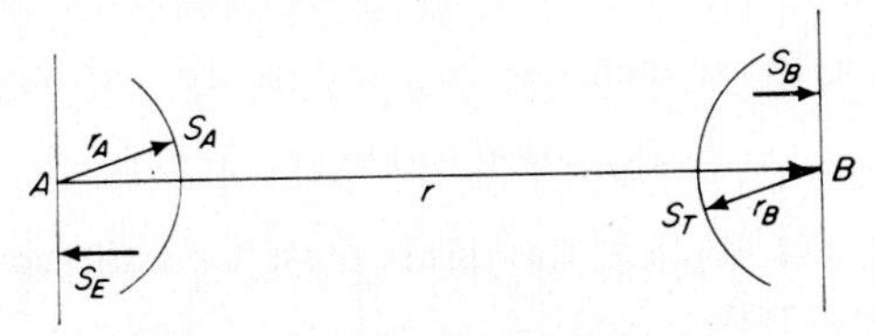

Fig. 4.1 Echo ranging diagram

We will start with absolute intensities to emphasize our model.

$$\frac{S_B}{S_A} = \left[\frac{r_A}{r}\right]^m e^{-\alpha' r} \tag{4.8}$$

The factor $e^{-\alpha' r}$ represents uniform absorption and the factor $\left[\frac{r_A}{r}\right]^m$ represents geometrical spreading.

$m=0$ means plane wave propagation
$m=1$ means cylindrical spreading
$m=2$ means spherical spreading

Typically, for ship sonar calculations, the propagation loss is described by a value of m between 1·5 and 1·9.

Allowing for the possibility that spreading on the return path may be different:

$$\frac{S_E}{S_T} = \left[\frac{r_B}{r}\right]^n e^{-\alpha' r} \tag{4.9}$$

Typically n varies from 0 to 2. (Refraction generally causes m and n to be functions of range, but we will ignore this for the moment.) Converting to decibels and putting $\alpha = 4{\cdot}3\ \alpha'$ gives

$$S_B = S_A + 10m \log r_A - 10m \log r - \alpha r \qquad (4.10)$$

$$S_E = S_T + 10n \log r_B - 10n \log r - \alpha r \qquad (4.11)$$

where α is the absorption coefficient in decibels per unit distance. Now recall that target strength is the ratio of reflected to incident intensity at the target, so by definition

$$T = S_T - S_B \qquad (4.12)$$

From Eqns. 4.10, 4.11 and 4.12

$$S_E = (S_A + 10m \log r_A) - [10(m+n) \log r + 2\alpha r] + (T + 10n \log r_B) \qquad (4.13)$$

Compare with Eqn. 4.1:

$$S_E = S_A - P_2 + T$$

Eqns. 4.1 and 4.13 are identical if r_A and r_B are both unity, and

$$P_2 = 10(m+n) \log r + 2\alpha r \qquad (4.14)$$

To make Eqn. 4.1 explicit, the terms must have the specific definitions given in the Appendix.

4.2.4 *Notes*

(i) *Source strength* So that Eqn. 4.13 will simplify to Eqn. 4.1 it is necessary that source strength (S_A) be defined as the intensity at unit distance from the source (i.e., $r_A = 1$). However, this is no use if the scale chosen is so small that the assumption of simple geometrical spreading is no longer valid. Both the sonar transducer and the target have diffraction effects which are characterized by a *far field* where our simple geometric spreading is valid and a *near field* where it is not valid. For typical sonar transducers the near field will extend to some centimetres or perhaps some metres from the radiating face. The near field of most targets is of greater extent, so unless the scale used is decametres or greater, a problem arises. The practical solution to this problem is to calculate the distance to the far field; measure the intensity in the far field, at known distance; then extrapolate this measured intensity back to unit distance from the source, using the far-field propagation law (in our case Eqns. 4.8 or 4.9). This gives the equivalent intensity referred to in the glossary. It then does not matter what units are used in the sonar equation, so long as they are consistent.

(ii) *The far field* The extent of the near field depends on the shape of the source and its type of oscillation; for instance a radially pulsating sphere has no near field for pressure measurements (though pressure is not in phase with particle velocity until $r \gg \lambda/2\pi$). However, for the conventional sonar it is safer to satisfy the far field criteria for a plane monopole radiator. To be certain that the measuring point is in the far field, the minimum measuring distance, r_{min} must satisfy three conditions:

$$\left.\begin{aligned} &\text{(a) } r_{min} \gg a \\ &\text{(b) } r_{min} \gg \lambda \\ &\text{(c) } r_{min} \gg \frac{\pi a^2}{\lambda} \end{aligned}\right\} \qquad (4.15)$$

where $2a$ is the maximum dimension of the source and λ is the sound wavelength. Translating these criteria into adequate practical inequalities for sea water:

$$\left.\begin{aligned} &\text{(a) } r_{min} \gg \frac{2}{f} \\ &\text{(b) } r_{min} \geqslant (2a)^2 f \end{aligned}\right\} \qquad (4.16)$$

where f is the frequency in kHz; r_{min} and a are in metres or yards.

(iii) *Target strength* To measure target strength one must pay attention to the same points as those discussed in Sections (i) and (ii). Sometimes target strength is defined as the scattering cross-section of the target divided by 4π, but this is precisely equivalent to our definition. The target strength will vary with the size and material of the target, with the orientation of the target and with the measuring frequency. It will also vary with the scattering direction, but we are only concerned with scattering back in the direction of incidence.

(iv) *Two-way propagation loss* (P_2) This is now the loss between the point at which S_A is measured, and the target; added to the loss between the point at which T is measured, and the receiver (given in our model, by Eqn. 4.14). The main point of note is that these losses are seldom the same (i.e., in general $m \neq n$). This can be illustrated with two extreme examples. Many textbooks treat the example of a sphere at relatively short range (but well into the far field) for which $m = n = 2$ and so

$$P_2 = 40 \log r + 2\alpha r \qquad (4.17)$$

Now if we replace the sphere with a large plane reflector, the two-way loss can be considered as the one-way loss for an image distant r behind the plane (see Fig. 4.2).

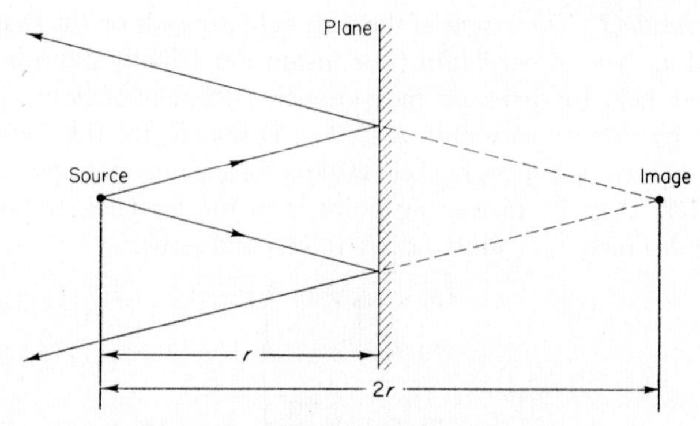

Fig. 4.2 Diagram of plane reflector

Hence

$$P_2 = 10m \log 2r + \alpha 2r$$
$$= 20 \log r + 6 + 2\alpha r \quad (4.18)$$

(at short range where $m=2$).

Comparing Eqns. 4.17 and 4.18 we see that the two-way spherical spreading of Eqn. 4.17 is replaced in Eqn. 4.18 by one-way spherical spreading plus a constant 6 dB loss on the return path, independent of r. Nevertheless, some authors calculate two-way propagation loss as spherical spreading both ways, which gives an error if the target is plane (for example the air/water boundary of the ocean). Thus, if we put $P_2 = 40 \log r + 2\alpha r$ in Eqn. 4.1, since we know $S_A - S_E = 20 \log 2r + 2\alpha r$ for a plane, then

$$T = 40 \log r - 20 \log 2r = 20 \log \frac{r}{2}$$

and the error in P_2 has been compensated by allowing T to be a function of r.

Eqn. 4.14 makes no allowance (except accidentally) for refraction. Refraction will generally be significant for medium and long range sonars, so P_2 needs to be measured or else calculated from a more detailed model than that leading to Eqn. 4.14.

4.2.5 Detection threshold

Now let us consider the meaning of the signal differential, R. Rearranging Eqn. 4.2 gives

$$R = S_{E(min)} - N$$

To understand what this means, we must consider what is meant by a 'just detectable signal'. Detection means making a decision. The receiver provides many output indications, some of which are signals, S_o, and some of which are noise, N_o. Decision is a very complicated and non-linear process. For instance, if one studies the receiver output indications based on some set of rules, eventually we must decide that the indication is YES or NO depending on how well it satisfies the rules. Once a decision is made, we are dismissing the contrary indications. For instance if we decide that the output does represent an echo despite its poor wave form, we are then treating it as if it had the correct wave form and dismissing the noise from consideration. The decision combinations based on output indications are shown in Table 4.1.

Table 4.1

Indication	Decision	Result	Probability
signal	signal	detection	p(D)
signal	noise	miss	$1-p$(D)
noise	noise	correct dismissal	$1-p$(F.A.)
noise	signal	false alarm	p(F.A.)

We must next ask what we want of our system. If it is essential that every possible detection is made, then this requirement is easily satisfied by calling every output indication a detection. In that way not many detections will be missed but a lot of false alarms will be caused. If the price to be paid for a false alarm is very high, then we could delay our decision indefinitely and have no false alarm, but alas—no detection either. It is natural to expect that the higher the ratio of signal indications to noise indications, the lower the probability of a false alarm for a given detection probability. Fig. 4.3 shows the relationship between the probabilities of detection and false alarm for various values of output signal to noise ratio marked on the curves in decibels. This calculation applies to Gaussian noise and a signal which is known completely except for phase. In practice reverberation noise and self noise are certainly not Gaussian, so a different optimum processing technique may apply (some people use a Rayleigh distribution for reverberation).

The idea of the price to be paid for one's choice of detection rules is an important one which is called the *cost*. One of the ramifications of cost can be seen in Fig. 4.3: for a given probability of detection, a low false alarm rate requires a high output signal to noise ratio, which in turn means a short detection range. We will use the symbol Δ for the output signal to noise ratio which corresponds to some stated value of p(D) and p(F.A.)

$$\Delta = S_o - N_o = \text{the number on the graphs of Fig. 4.3.}$$

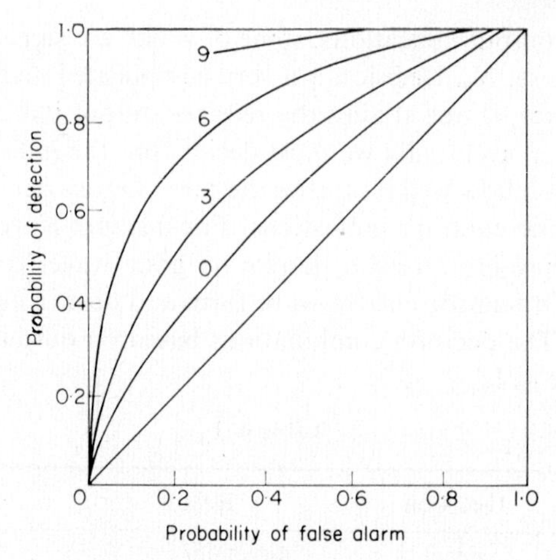

Fig. 4.3 Detection performance

Fig. 4.4 shows that signal processing allows detection systems to have a gain which we will call G, so the value of input signal to noise ratio $(S_i - N_i)$ corresponding to Δ is $(\Delta - G)$.

In practice G is a complicated function of Δ and of the system parameters; for instance notice that the gain increases with the product of time τ and bandwidth b of the echo and that for small signals, the output signal to

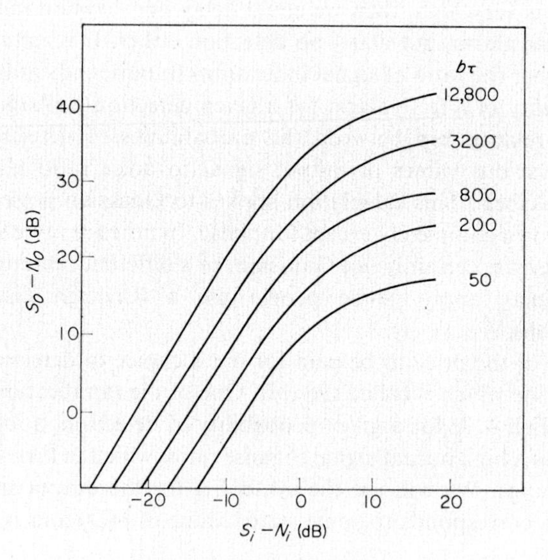

Fig. 4.4 Processing gain

noise ratio is the square of the input signal to noise ratio. To remind us that G is a function of Δ, we will denote the input signal to noise ratio corresponding to some stated output signal to noise ratio as $(\Delta - G_\Delta)$. This input signal to noise ratio is called the *detection threshold* of the processing system. Note that detection threshold is not a fundamental property of the system, but a result of the decision rules applied at the output.

Now return to the sonar equations, to make them explicit in the light of this discussion. We have

$$S_o - N_o = \Delta \qquad \text{(definition of } \Delta)$$

$$S_i - N_i = \Delta - G_\Delta \qquad \text{(definition of } G_\Delta)$$

$$\therefore \quad S_i = N_i + \Delta - G_\Delta \qquad †(4.19)$$

and from Eqns. 4.2 and 4.3

$$S_E = N + R + S_x \qquad (4.20)$$

S_i corresponds to S_E; N_i to N so

$$\Delta - G_\Delta = R + S_x$$

giving us the alternative forms of the sonar equations

$$\mathbf{S_A - N - \Delta + G_\Delta - P_2 + T = 0} \qquad (4.21)$$

$$\mathbf{S_p - N - \Delta + G_\Delta - P_1 = 0} \qquad (4.22)$$

Here we notice that the vaguely defined recognition differential plus the nebulous excess signal level now become explicitly the detection threshold. R is sometimes defined as the input signal to noise ratio for a 50 per cent probability of detection, but there is a difficulty in this definition: Δ and G_Δ are real parameters, even though very complicated ones, but they only define $(R + S_x)$.

The remainder of the explicit definitions for the terms of the sonar equations as shown in the Appendix, now follow. An important point to note is that all these conditions must hold at once before the sonar equations can be used in the forms of Eqns. 4.21 and 4.22.

4.2.6 *General comments on the sonar equations*

The sonar equations are an attempt to represent a very complex situation by some factors which can be measured. Unfortunately the terms in the so-called equations are not independent, so attempts to investigate parts of the equation often come to grief. We have already seen that G is a function of

† This whole expression is more a definition than an equation. For instance $2(S_i - N_i) = 2\Delta - G'_{2\Delta}$ but $\neq 2\Delta - 2G_\Delta$.

Δ and P is a function of the characteristics of the source and target as well as the medium. T is not only a characteristic of the target, but sometimes a function of range. For instance the target strength of a wake for a narrow beam increase with range because the greater the range, the more of the wake intercepts the incident sound, so the scattering cross-section increases. Also, T decreases for a very short duration ping. The term N is a most complicated one, and one of its components—reverberation—is a function of source strength, S_A and details of the signal processing. All this adds up to the need for caution in looking at just parts of the equation. For instance the Figure of Merit for a sonar set which is functioning correctly may vary 30 dB as a function of features external to the set, such as plausible combinations of target velocity, observer velocity, sea state and detection threshold setting.

Last, but by no means least, if there is a human being in the decision process, the complexity of describing the overall performance increases enormously. We could hardly expect an operator's performance to be optimum when he is sea sick, tired or absent. This is sometimes allowed for by an extra term in the sonar equation, which represents 'operational degradation'. Not so apparent are the psychological factors. Human beings have become accustomed to an irritating process known as learning, so a spell of investigating false alarms is certain to cause a rise of detection threshold. On the other hand, independent evidence of the presence of a target (such as would be provided by a torpedoed ship) causes a great reduction of detection threshold. Consequently there is ample scope for psychologists in studying detection.

4.3 Signal processing

Fig. 4.5 shows a section of an echo sounder chart with two target tracks on it, echoes from the shallower target being much stronger than those from the deeper one. The signal to noise ratio of the echoes remains constant, but the bias has been changed, which corresponds (to a certain extent) to a change of detection threshold. In the top and bottom settings the detection threshold is suitable and both target tracks can be seen. In the second band the detection threshold has been set so high that only the very strong echo marks the chart. This is unsatisfactory because the depth might be uncomfortably shallow before the chart marks. The third band shows a detection threshold which is set so low that scatterers at all depths are marking strongly, masking the weak echo.

Consider the echoes from a single ping (Fig. 4.6(a)) and try to make a decision on the depth of the target. Without considering cost, one would choose the most intense echo marked by the arrow.

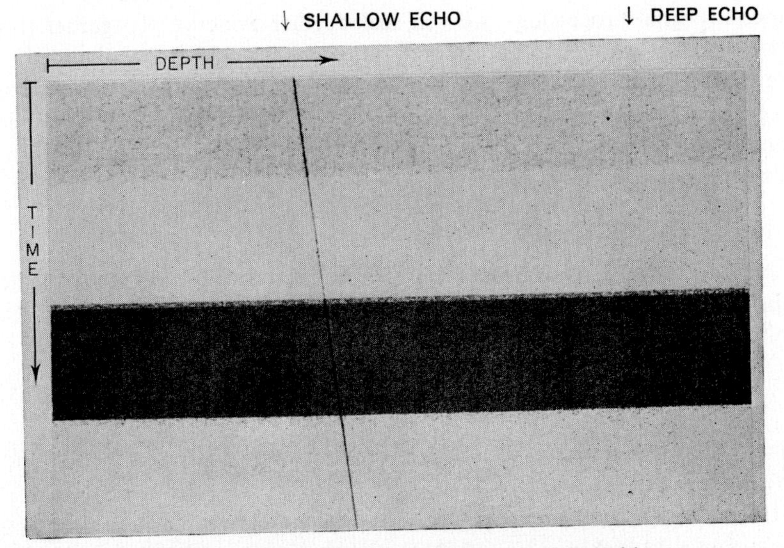

Fig. 4.5 Demonstration of detection threshold

The owner of the ship, however, might weight the evidence with a taper to the right, Fig. 4.6(b), on the grounds that a missed detection at very short range would cost him more than a miss at extreme range. The weighting applied to the evidence would then determine which echo should be chosen. This is what was referred to earlier as a set of decision rules. To try and resolve this, we will now do something of the greatest fundamental impor-

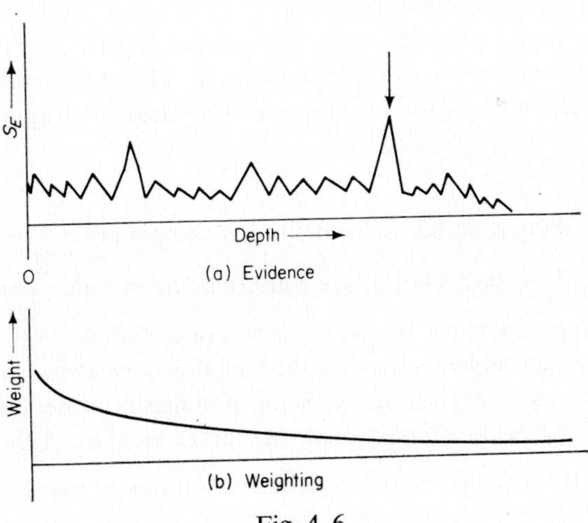

Fig. 4.6

tance to signal processing—we will seek further evidence. To gather this extra evidence we send out additional pings and the echoes are all added to the chart. Sooner or later, we become confident that we have located the target simply from our knowledge that a real target would only change depth systematically. This increase of confidence is caused by visual integration which will be demonstrated below.

4.3.1 Processing gain, G

Now to try to explain processing gain in a few words we start with the relation for combining pressures p_1 and p_2. For two contributing sample points

$$\text{Total intensity} = \frac{1}{\rho c}\left[\overline{(p_1+p_2)^2}\right] = \frac{1}{\rho c}\left[\overline{p_1^2}+\overline{p_2^2}+2\overline{p_1 . p_2}\right] \qquad (4.23)$$

where the bars indicate expected averaging and ρ=density; c=sound velocity.

If the two pressures p_1 and p_2 are statistically independent (i.e., p_1 and p_2 are uncorrelated) then $\overline{p_1 . p_2}=0$. Now, combining two identical signals, q, we get

$$\text{Signal intensity} = \frac{1}{\rho c}\left[\overline{q^2}+\overline{q^2}+2\overline{q . q}\right] = \frac{4\overline{q^2}}{\rho c}$$

and combining two equal power noises we get

$$\text{Noise intensity} = \frac{1}{\rho c}\left[\overline{n^2}+\overline{n^2}+0\right] = \frac{2\overline{n^2}}{\rho c}$$

(Covariance of random noise is zero.)

Thus signal intensity increases with the square of the number of contributing points, but noise intensity increases linearly with the number of points, so the signal to noise ratio for intensity increases in proportion to the number of contributing points.

Therefore

$$G = 10 \log \text{(number of contributing sample points)} + \ldots \qquad (4.24)$$

The restrictions on this relation are only that the sample points must be:

(a) far enough apart for the noise to be uncorrelated;
(b) close enough together so that the signal is coherent;
(c) valid, i.e., the target echo must be potentially present all the time (e.g., if the beam is directional, the target must be in the beam).

In practice the evidence may be gathered at different places at the same time, in which case it is called array gain; or else it may be gathered at one

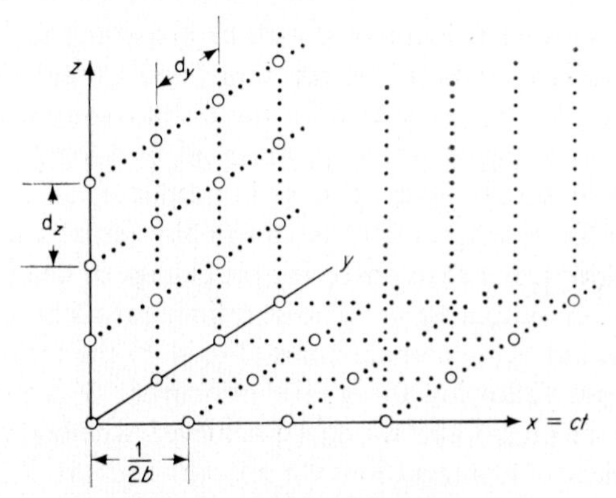

Fig. 4.7 Volume of signal wave

place at different times and integrated, i.e., evidence may be gathered across or along the wavefront. This can be visualized as the *volume* of the signal wave sampled, provided that you can visualize a 4-dimensional volume. Fig. 4.7 demonstrates this in 3 dimensions only.

The spacing of the hydrophones in the yz plane (the array) must satisfy condition (a) and (b) above and minimum spacings d_y and d_z must be determined by experiment. Typically (a) and (b) are not satisfied completely and so there is a loss of gain, see Fig. 4.8.

In the $x=ct$ dimension, sampling theory tells us that the minimum spacing in order to satisfy conditions (a) and (b) above is $1/2b$ where b is the bandwidth of the processing equipment. Therefore, in a total time τ along the t axis there are $2b\tau$ sample points. So

$$G = 10 \log 2b\tau + 10 \log (\text{number of hydrophones}) + \ldots \qquad (4.25)$$

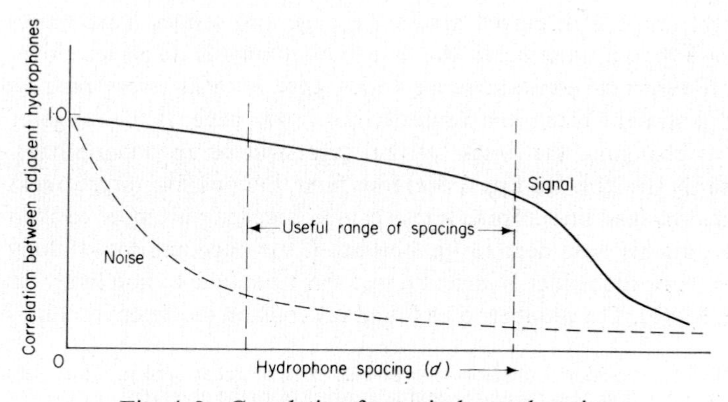

Fig. 4.8 Correlation for typical sample points

Recall Fig. 4.4 showed G increasing with $b\tau$ according to Eqn. 4.25. As there is a square law relation between input and output signal to noise ratio, if the $10 \log 2b\tau$ is appreciated on the display (output), then this is equivalent to a lowering of detection threshold by $5 \log 2b\tau$. This can be seen in Fig. 4.4 by observing the change in output signal to noise ratio as a function of $b\tau$ for constant input signal to noise ratio (equivalent to pre-detector integration), and then observing the change of detection threshold with $b\tau$ for constant output signal to noise ratio (equivalent to post-detector integration). On our echo sounder chart (Fig. 4.5) the evidence increased rather slowly, but following the above principles, G is proportional to 10 log (number of pings) and would be achieved whether we used visual integration or electronic integration. On this sort of chart, this is equivalent to saying that the gain is proportional to the total area of target echo present. A similar rule is quoted by Tucker[2] for a very small pip on a noisy PPI† display. Visual integration has not yet been accurately accounted for mathematically. Some displays give the gain to be expected from our simple argument, but a display like Fig. 4.5 produces a reduction of detection threshold of about $8 \log 2b\tau$.

A proper evaluation of processing gain is given by Stocklin[3] for four types of processing, based directly on the statistical properties of the basic acoustic field.

Visual integration Eqn. 4.25 springs to life with a real-time demonstration of visual integration. Such a demonstration is easily captured on cine-film, but it is not so easy to be convincing in a limited space with stills. Nevertheless, if the reader approaches our demonstration with a friendly attitude, we may succeed. Fig. 4.9(a) and (b) represent two different sections of an intensity modulated range recorder or echo sounder chart, like Fig. 4.5.

Target range is displayed across the page and echoes from successive pings are drawn underneath the first so that time is displayed down the page. A target at constant range which gave an echo every ping would 'paint' a straight black line vertically down our page. If the range of the target is changing, the 'track' of the target will be an oblique line—not necessarily straight. On Fig. 4.9(a) four faint, but possible target tracks are indicated by lead lines. These tracks can be 'smelled out' more easily if the reader puts his nose near to the bottom of the page and looks along the arrows. Now the reader is invited to use this technique to find likely targets in Fig. 4.9(b). The number to be found depends on the detection threshold

† PPI = Plan Position Indicator—the normal sort of radar screen, which displays targets as a function of range and azimuthal bearing from the observer—usually with the observer's position represented at the centre of a cathode ray tube screen.

Fig. 4.9 Demonstration of visual integration

chosen, that is, how hard you try. Now Figs. 4.10(a) and (b) should be similarly studied. Most (normal?) readers should now be confused about the location of targets. The demonstration of visual integration really consists of joining Figs. 4.9 and 4.10 together and looking at all the pings at once. This has been done in Fig. 4.11, and the one consistent target track is marked in with lead arrows. This track can now be readily identified in Figs. 4.9 and 4.10.

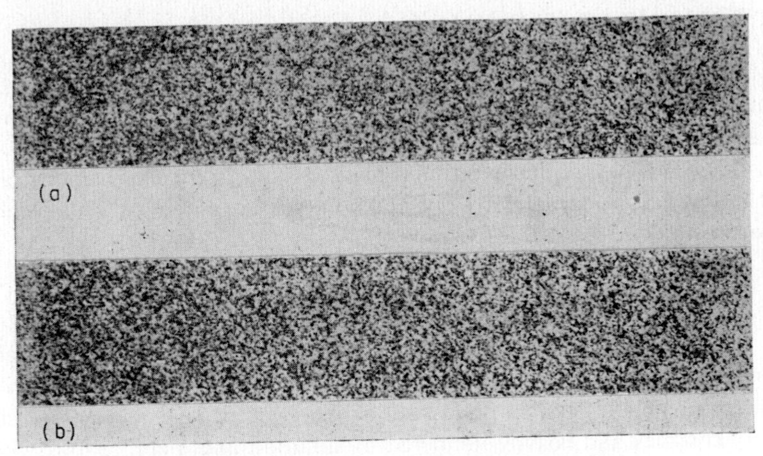

Fig. 4.10 Demonstration of visual integration

A slightly more realistic demonstration can be obtained from Fig. 4.11 alone. Cover the chart with a card and also hide the lead arrows. Lower the card smoothly, exposing the chart uniformly with time. Usually, more than half the chart must be exposed before one can have any confidence about any particular track, but as the evidence accumulates, one becomes more

Fig. 4.11 Demonstration of visual integration—composite record

and more confident that the marked track is the correct one. This same pattern of behaviour is observed if the experiment is repeated with the book upside down so that the detection is made back to front. This demonstrates that the effect is a progressive one and not a feature of the particular echoes portrayed.

4.3.2 Other parameters

Processing gain, G, is not the only important consideration in signal processing; it is easy to think of other vital facets of decision-making. Unfortunately a detailed discussion of signal processing rapidly becomes very complex and heavily lumbered by notation and qualifications which would be out of place in this non-rigorous review. Instead we will just mention some of the ideas which are important. Siebert[4] discusses four measures of performance quality:

(a) the *reliability* of detection,
(b) the *accuracy* with which target parameters can be estimated,
(c) the extent to which such estimates can be made without *ambiguity*, and
(d) the degree to which two or more different target echoes can be separated or *resolved.*

(i) *Reliability* We have already touched on this subject in Section 4.2.5 where we noted that in order to achieve a specified reliability of detection, it is necessary to have a great enough signal echo to noise ratio. On close scrutiny it turns out that the pertinent ratio is an energy ratio and this condition is almost independent of the waveform of the ping. This is a good point to introduce the *energy principle* which is stated simply by Stewart and Westerfield[5] thus:

> 'The energy principle: As far as Gaussian noise is concerned, the choice of signal waveform and bandwidth is entirely arbitrary. One can send out the same energy in a long pulse of low power or a short pulse of high power and obtain the same [*reliability*] result.'

This is a particularly important result for underwater acoustics, because cavitation at the source sets an upper limit to the source strength, S_A. The sonar equations 4.4, 4.21 give no clue how to improve the reliability of detection for fixed source strength and noise, but the energy principle says that reliability will increase with the duration of the ping. This improvement with ping duration is contained in the factor G (processing gain) and was mentioned in Section 4.3.1 (Eqn. 4.25). However, if the background noise is reverberation, this improvement with pulse duration does not go on indefinitely—there is an optimum pulse duration for the type of target.

(ii) *Accuracy, ambiguity and resolution* The main target parameters which can be obtained from a sonar set are:
(a) range, given by the echo delay time,
(b) bearing, given by path difference measurements to different parts of the receiving transducer array, and
(c) relative velocity, given by Doppler frequency shift of the echo.

Accuracy in measuring these quantities is wanted so that the observer will know *where* the target is. Good resolution in range and bearing is wanted so that the *shape* of a target can be determined, and good Doppler resolution is wanted so that slow moving targets can be *identified* from stationary ones.

These are important aids in target classification. The accuracy and resolution obtainable in measuring these target parameters depends on the waveform of the ping. Suppose that we are using a CW ping and the prime requirement is high reliability of detection (or extreme range of detection for stated reliability). The energy principle indicates that a ping of very long duration should be used, and the echo should be processed through an 'optimum' filter of bandwidth equal to the reciprocal of the ping duration.† So a ping of long duration indicates a narrow filter bandwidth, which allows another filter of the same bandwidth to be used next to the first and thus *resolve* a very small change of frequency, corresponding to a small target velocity. However, the output of a very narrow band filter changes so slowly that the instant of arrival of an echo cannot be determined very precisely, so although a long CW ping has a long detection range and good Doppler resolution it has poor accuracy in measuring target range. Even if a filter of wide bandwidth were used, the range resolution would be poor, because a long duration CW pulse occupies a long stretch of water with consequent position ambiguities. Conversely, a short duration CW ping is characterized by short detection range and poor Doppler resolution but good range resolution and accuracy.

These quality measures which depend on the waveform of the ping are linked by an uncertainty principle which is discussed by Siebert[4]; Stewart and Westerfield[5] and Federici[6]. This is easy to believe when we note that evaluation of the parameters (a), (b) and (c) above involves measurements of echo time delay, τ and phase, ϕ. Stewart and Westerfield[5] use the name *signal ambiguity* for the uncertainty of position of the target in the τ, ϕ domain.

(iii) *Data rate* This is an additional measure of performance quality. A high rate of information is particularly important if the observer or target velocity is very high, or if the propagation conditions are changing rapidly. For instance, if the interval between pings is so great that the target can change position significantly between pings, then the potential processing gain will be reduced because the systematic motion of the target (which is necessary for visual integration) is absent. Unfortunately, a sonar which is designed for long detection range will have a large echo delay and so an inherently low data rate. In such a case it would be necessary to interleave other coded transmissions to keep the data rate to an acceptable figure. In simpler sonars it is sometimes necessary to sacrifice long detection range for adequate data rate, as for example where sonar is used for navigational data for a fast vehicle.

† In practice, a bandwidth a few times greater than this causes only a small drop in reliability. For some applications, such a wider filter bandwidth is required to accommodate Doppler shift—as for example when detection is made aurally.

4.4 Background noise

4.4.1 Types of noise

In the *sonar equations*, the masking background noise intensity against which the receiver must detect the echo is given by

$$N = 10 \log (n_r + n_s + n_a)$$

where n_r = reverberation intensity
n_s = self noise intensity which is made up of two contributions:
(i) electrical receiver noise and electrical interference
(ii) acoustical self noise from own machinery and motion effects
n_a = ambient sea noise intensity

Note: each noise term is the intensity in the bandwidth of the receiver channel being considered. Each of the noise terms has a frequency dependence and they are all quite different.

4.4.2 Reverberation

This is energy from the ping which returns to the receiver from scatterers other than the target. It therefore has somewhat the nature of a target echo. In fact some scatterers produce echoes which are so similar to target echoes that they are called 'false targets' or 'non submarines'. The greatest single problem of present day and future active sonars is not that of obtaining an echo from a target at long range, but rather of deciding which of the *many* scatterers within the range of surveillance is likely to be a target and how likely? An understanding of the nature of reverberation is of great importance for modern exploration.

Those who study reverberation, identify three main sources; surface reverberation, bottom reverberation and volume reverberation. Fig. 4.12 shows how the echo level might vary immediately after transmissions in deep and shallow water, with indications of the predominant cause of the reverberation in each case.

Surface reverberation is discussed fully in Chapter 3 as back scatter. Bottom reverberation usually predominates in shallow water, or when seeking targets on the bottom and volume reverberation is sometimes characterized by well defined scattering layers, such as the one marked SSL in Fig. 4.12. Some very interesting results of the scattering properties of fish are presented by Brekhovskikh[7].

Reverberation is too difficult to try to discuss in the space available. Instead some results of practical interest will be listed.

(i) *Dependence on sea state* In deep water one might expect reverberation to depend on the roughness of the sea surface. It is found that the reverberation immediately after a ping is markedly affected by the surface roughness, but the dependence of reverberation on sea state falls with time from ping, which leads to the rather surprising result that reverberation from beyond 100 yards is practically independent of sea state.

(ii) *Wakes as false targets* Scattering from the small bubbles left in the wakes of ships is quite pronounced. Wakes have often been mistaken for targets so the effect has been studied in some detail.

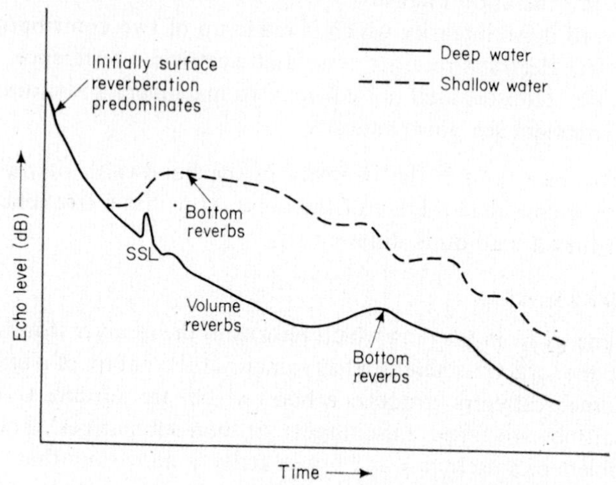

Fig. 4.12 Reverberation in deep and shallow water

(iii) *The deep scattering layer* Twenty years ago a sonic scattering layer (SSL) was frequently detected by sonar sets in the 15–30 kHz frequency band at depths from 50–400 fathoms in the daytime. This layer was generally believed to be of biological origin because it migrated to the surface at night, then descended again in the morning, but in those days, attempts to identify the organism responsible (by trawling) were always unsuccessful, so it was assumed that the efficiency of the catching nets was low. Indeed, the dominant organism has not yet been identified beyond all doubt, but much more evidence has accumulated in the last two years.† Experiments using wide-band sources have shown that there are large peaks in the curve of scattering cross-section versus frequency and that the frequencies of the maxima change as the layer changes depth. This behaviour is consistent with the hypothesis that the peaks represent resonant scattering from the gas-filled swim bladders of fish. If this is so, the actual concentration of scatterers necessary to explain the measured scattering cross section is very much smaller than was originally thought and is consistent

† 1966.

with concentrations deduced from net catches[8]. Barham[9] has made direct visual observations near California in conjunction with scattering measurements at 12 kHz. He finds a correlation between the scattering layer and the concentrations of (a) small bladder-fish (myctophids) and (b) transparent jelly fish with a swim bladder (siphonophores).

In daytime the scattering strength of the SSL increases monotonically between 2 and 12 kHz. At night, scattering strengths are higher and nearly uniform between 3 and 12 kHz with a peak at 8 kHz. The dominant cause of the SSL probably varies with frequency, possibly as follows:

(a) 4–6 kHz strong resonant scattering from bathypelagic fish—mainly myctophids,
(b) 6–15 kHz myctophids and siphonophores,
(c) 15–40 kHz resonant scattering from siphonophores and larval cyclothone,
(d) above 40 kHz—non-resonant scattering—(euphausiids, etc.).

(iv) *Transducer beamwidth effects* There are beamwidth effects of importance to reverberation. A strong sidelobe can gather reverberation from a direction quite different from the axis of the main beam. Thus surface or bottom reverberation can be of importance even with a transducer pointed at a target at mid depth. Another important effect of finite beamwidth is change of reverberation spectrum.

Fig. 4.13 shows a ship travelling at speed v with the main axis of its sonar

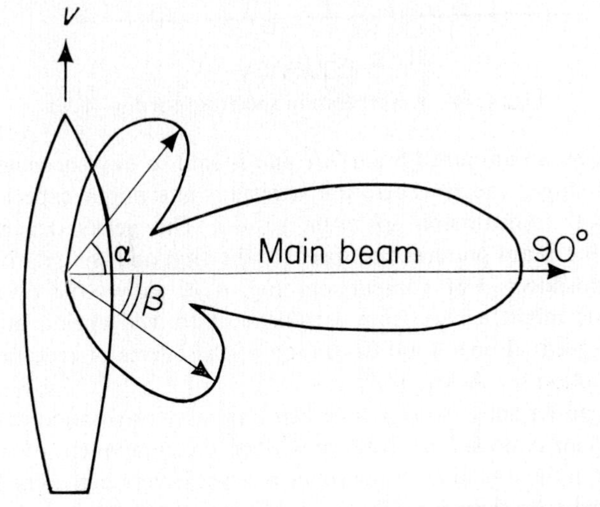

Fig. 4.13 Diagram showing main beam and first side lobes

transducer pointing sideways at what is called a relative bearing of 090°. There is no relative speed between the receiver and scatterers on the axis of the main beam. However, the ship is approaching scatterers at the angle marked α with a speed $v \sin \alpha$, so reverberation from these scatterers will be raised in frequency (f) because of the Doppler effect by an amount $(fv \sin \alpha)/c$. Reverberation from this direction is said to have *closing Doppler*. Similarly reverberation from the direction β will be lowered in

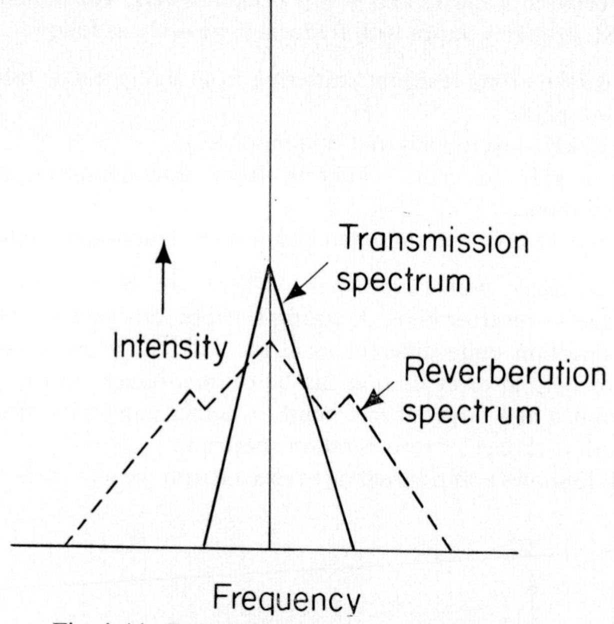

Fig. 4.14 Reverberation spectrum for Fig. 4.13

frequency by an amount $(fv \sin \beta)/c$ and is said to have *opening Doppler*. Fig. 4.14 shows the reverberation spectrum one might expect from the hypothetical transmission spectrum shown. The peaks depend on the shape of the beam pattern, the speed of the ship and the relative bearing of the transducer. For comparison, Fig. 4.15 shows the reverberation spectra one might expect from a transducer trained ahead, at high and low ship speeds. For a good discussion of the effects of reverberation on sonar detection see Ackroyd[10].

In high power active sonars, reverberation noise predominates except for special circumstances. One instance is when the component of the target's real speed in the direction of the sonar receiver is very high. The frequency of the target echo then has a Doppler shift away from the reverberation, so filtering can be used to select the echo frequency. This target Doppler generally is not much help in improving the signal to noise ratio in geological or biological exploration, but is useful for detection of vehicles. Moreover, the *amount* of the echo frequency shift can tell the observer the *value* of the target speed in his direction. One difficulty is that the Doppler frequency shift depends on the component of the target's real speed in the direction of the observer and also on the component of the observer's real

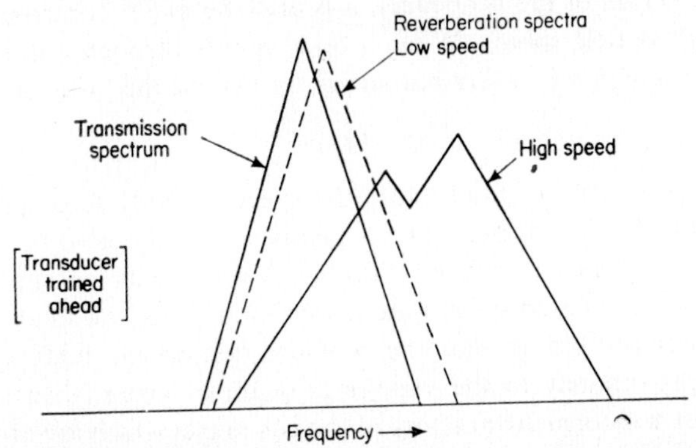

Fig. 4.15 Reverberation spectra at high and low speed

speed in the direction of the target. To solve this we notice that the reverberation comes from stationary scatterers and so *its* Doppler shift depends only on the observer's speed. Therefore we can eliminate the effect of the observer's motion by comparing the frequency of the target echo with the centre frequency of the reverberation (if the target is on the axis of the beam). Regarding reverberation then as a target of zero Doppler, for a 10 kHz sonar set, a real target will show about 6·8 Hz frequency shift for every knot.

A second occasion when reverberation does not predominate is when the self noise is made abnormally high by such common causes as:

(a) very rough seas,
(b) very high vehicle speed, or
(c) machinery defects causing excessive electrical or mechanical interference.

4.4.3 *Self noise*

A sonar set is usually designed so that electrical self noise is lower than ambient sea noise. Therefore high electrical self noise is usually an indication of a defect, such as electrical interference, an electronic fault or loss of transducer sensitivity. The bulk of self noise is usually acoustical in origin and comes from the ship's own machinery and from the effects of motion.

Vibrations from machinery in a ship can be transferred through the hull to the sea and also directly to the transducer. Some of the radiated noise also is picked up in the transducer as self noise.

(i) *Cavitation noise* At speeds greater than the normal cruising speed, the major contributor to *radiated* noise from most ships is propeller cavitation noise. Unless very special precautions are taken to exclude noise from the direction of the propellers, it is also the major contributor to *self* noise at high vehicle speeds. When a body moves through a fluid of static pressure p, there is a pressure reduction Δp behind the body, given by

$$\Delta p = C(\tfrac{1}{2}\rho u^2) \tag{4.26}$$

where ρ is the density of fluid and u the speed of body through the fluid. The constant of proportionality C, depends on the shape of the body and is called the pressure coefficient. Pressure coefficients or similar shape coefficients are tabulated for common shapes in books on fluid dynamics. If Δp is great enough so that the pressure behind the body $(p - \Delta p)$ is reduced approximately to the vapour pressure of water, then cavities of water vapour will form around any of the gas nuclei which are abundant in sea water. As these cavities are swept into regions of normal pressure p, they will collapse violently with the emission of acoustic noise. This process occurs fairly randomly, so noise is emitted over a wide bandwidth. A detailed description of the dynamics of growth and collapse of these transient vaporous cavities is very complicated and unnecessary for our discussion. We will call the actual pressure at which cavitation noise appears the cavitation pressure p_c. In sea water it is very nearly equal to the vapour pressure. Actually dissolved air will start to come out of solution into large nuclei at moderate pressure reductions below static (i.e., depends on Δp, not $p - \Delta p$), but this gaseous cavitation is an acoustically gentle effect which is not significant in the radiated noise from ships.

Consider the cigar-shaped vehicle (Fig. 4.16). If it is a good shape, the

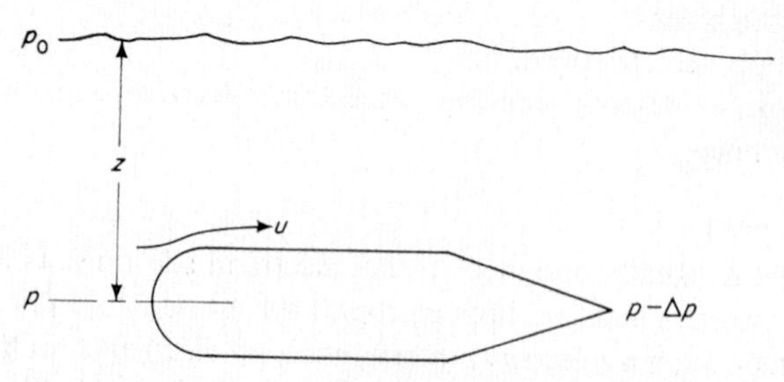

Fig. 4.16 Diagram of body moving through the sea

body might cavitate in the range 30–40 knots; but if there is a rough object, such as a bolt head protruding from the side of this vehicle, the rough object might have a very large coefficient of pressure, so cavitation will occur behind the protruberance at a much lower speed, say 15 knots. This type of cavitation is most serious as a cause of self noise when it occurs at some point on the hull or on the sonar dome which is very close to the transducer. Consequently for high speed operation great care is needed to design a dome with a low coefficient of pressure and to smooth off or fair any protruberances in the vicinity. Even so, this type of self noise usually sets the upper speed limit for satisfactory operation of sonars.

Now consider a particular propeller of outside diameter d, turning in water at a depth z. Here the value of fluid velocity, u, at a point on the blade depends on the rate of turning $\dot{\theta}$ and the radius, x, of that point. At the tip of the propeller, where $x=d/2$, u will be much greater than v, the speed of the vehicle, so we can expect that the tip of the propeller will cavitate at a much lower vehicle speed than does the hull. We will rephrase Eqn. 4.26 in a form suitable for considering propeller cavitation

$$p_c - p_0 = \rho g z - C.\tfrac{1}{2}(x\dot{\theta})^2 \tag{4.27}$$

Where p_0 is the static pressure at the surface (1 atmosphere) and g is the gravitational acceleration. So cavitation will occur for the condition

$$p_c - p_0 \geqslant \rho g z - C.\tfrac{1}{2}(x\dot{\theta})^2 \tag{4.28}$$

Consider first just the tip of the propeller, for which x is a constant ($=d/2$). From Eqn. 4.28, cavitation will only occur for

$$z \leqslant K_1 \dot{\theta}_c^2 + K_2 \tag{4.29}$$

K_1, K_2 are constants. The rate of revolution for the *onset* of cavitation at the propeller tip at depth z is found by equating 4.29. This shows that the deeper the propeller, the faster the vehicle can go before the onset of cavitation. This is why modern submarines travel very deep so that they can make use of high speed without causing undue self noise and without declaring their presence by high radiated cavitation noise.

It is possible to increase the cavitation inception speed of a ship considerably by designing a propeller with a low value of C. The pressure reduction is of course essentially tied up with the lift which provides the thrust for propelling the ship, but better performance is achieved by reducing the lift towards the tip of the blade so that the pressure reduction is uniform over a fairly large area of the blade. Increasing the area of the blades also gives a smaller pressure reduction for a given thrust. However, it is not possible to make all blades identical, and in any case, the slightest amount of wear or damage will cause the different blades to have different values of C. Thus the different blades of the propellers have different cavitation inception speeds. As a submarine comes up from the deep at constant propeller revolutions we get this pattern.

(a) Initially there is no cavitation.
(b) Cavitation starts on the tip of one blade of one propeller. The depth is given by Eqn. 4.27 in which $x=d/2$.
(c) As the depth decreases slightly, tip cavitation starts on other blades which have a lower pressure coefficient. (See Fig. 4.17.)
(d) At shallower depths the tips cavitate more violently and the cavitation spreads down the trailing edge of the blade (to lower values of x).
(e) At even shallower depths, the whole of the trailing edge cavitates and part of the leading surface does too. (See Fig. 4.18.)
(f) If the submarine is travelling fast enough, hull protruberances start to cavitate.

Fig. 4.17 Propeller showing tip cavitation

Now one last point before we leave propeller cavitation. Note that the tip of each blade changes depth during each revolution by an amount *d*, the diameter of the propeller. Therefore the *onset* of cavitation will occur on the worst blade, only at the top of each revolution. This makes it much easier to

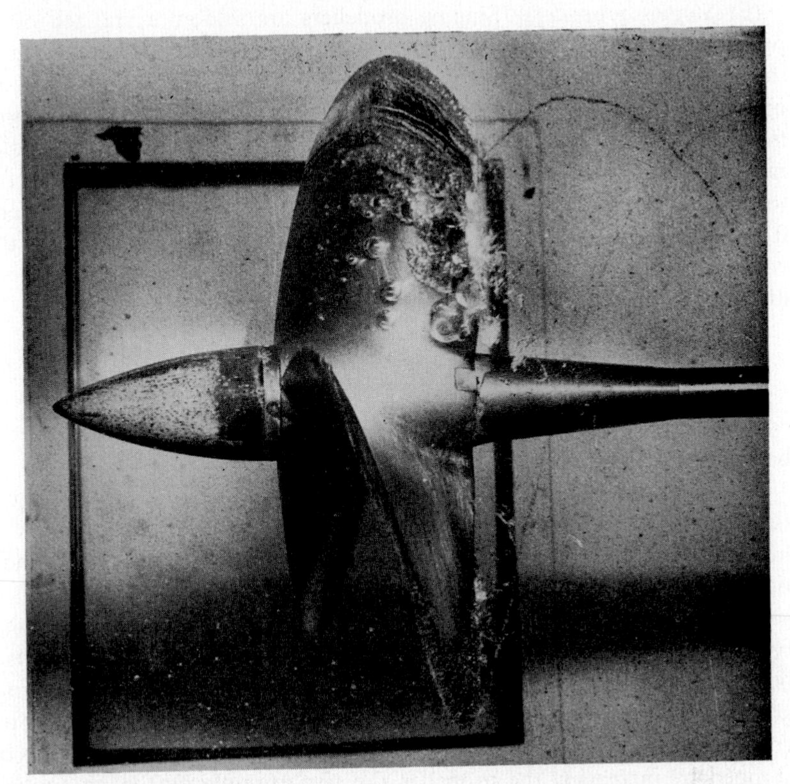

Fig. 4.18 Propeller cavitation has spread down the blades

hear, because the cavitation noise will be intensity modulated at the shaft rate.† When *all* the blades are cavitating, *each* will be intensity modulated at the shaft rate, so the radiated noise from each propeller will be modulated at the blade rate which is equal to the shaft rate multiplied by the number of blades. This too is easy to count, so by listening to the sound radiated by cavitating propellers, one can learn the shaft rate, the number

† It is a fact that it is easy to hear the modulation of the cavitation of each blade at the shaft rate, but the reason may not be simply the change of static pressure at the blade tip. It is more likely to be a variation in p_c as the blade moves into different parts of the flow pattern.

of blades and the number of propellers. This can tell a listener quite a lot about the vessel making the noise. A vessel running near cavitation inception speed often shows a marked variation of radiated noise as the screws change depth in a swell.

(ii) *Singing propellers* Singing propellers are seldom a real self noise problem, because of the low frequency and relatively narrow bandwidth of the tone. They do contribute handsomely though to the radiated noise of many vessels, so we will discuss them here for tidiness. The exciting force for this noise is the Karman vortex street which is shed behind bodies moving in a fluid. Examples usually quoted are the Aeolian Harp, and the tone from the telephone wires in a strong wind. Other examples are the tone from the slip stream on a car roof rack when it is empty, and probably the whistle from an aerial bomb. The behaviour is characterized by a non-dimensional constant K called the Strouhal number.

$$K = \frac{fd}{u} \tag{4.30}$$

where f is the frequency of the vortex tone and d is the diameter of the body.

Nothing much happens until the frequency of the exciting vortex is equal to that of a mechanical resonance in the solid body or until a condition of high coupling of the vibration to the water occurs. Under the right conditions a very high intensity tone can be radiated from the structure. This is sometimes called hydroelastic noise[11]. Propellers of ships are prone to this type of radiation. The meaning of fluid velocity u for a singing propeller is a little vague, since the radiation starts abruptly at a given value of shaft rate and thrust but then continues at the same frequency as the shaft rate is increased over a fair range. At a given shaft rate the singing can sometimes be stopped by increasing the propeller slip, for example by towing another ship. This probably has the effect of cutting down the correlation length along the blade. Two other examples of excitation by Karman vortices are of interest in underwater acoustics. One is when a sonar dome is excited enough to rattle and cause local self noise to hinder low frequency listening. The other is the familiar strumming of anchor cables in the current. If a hydrophone is lowered on a cable in an ocean current this strumming can be communicated to the hydrophone.

(iii) *Flow noise* Sonar men used to speak of flow noise as a self noise from the turbulent flow of water past their hydrophones. Flow noise was said to increase with the speed of the flow and the roughness of the surface of the dome. Not much was known of the subject in the sonar world. Now hydrodynamicists have specific meanings attached to various components of flow noise and the whole subject is very complicated and not yet fully resolved[12] so we will leave it alone. We will just note that in turbulent flow past a very small hydrophone there are pressure fluctuations carried along in the flow which would be converted to voltages. On a large hydrophone the pressure fluctuations would average out. This is called pseudo-sound, or non-acoustic noise, because it is only detected by a hydrophone when the flow is in direct contact with the hydrophone face: it does not propagate like normal sound. There is also a component of true acoustic sound associated with turbulence which is radiated out of the near field—but it is of very low level.

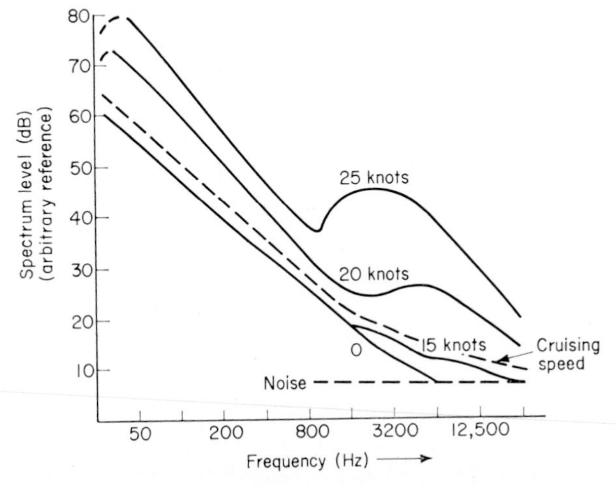

Fig. 4.19 Ambient noise in sonar dome

The roughness of the surface has an influence on the turbulence. Skudrzyk and Haddle[13] found that there is a threshold size, h, for the roughness before it will significantly affect the pseudo-sound level:

$$h\ (\text{inches}) = 6 \times 10^{-2} u^{-1}\ (\text{knots}) \qquad (4.31)$$

Experimentally, it is easy to confuse flow noise with cavitation noise—a true acoustic noise which also increases with flow rate and surface roughness. Cavitation sets in abruptly at a threshold flow velocity, while flow noise is present to some extent at all speeds.

(iv) *Machinery noise* Fig. 4.19 shows equivalent isotropic noise in a sonar dome. The noise is practically unchanged from 0 to 15 knots, so we can deduce that auxiliary machinery is the main cause of self noise at low speed. Noise caused by the main engines starts to predominate above cruising speed (cruising speed of a ship is usually an economical speed at the knee of the drag curve and is determined by wave-making parameters such as length of ship and type of hull). We also see that above cruising speed the high frequency noise rises more rapidly with ship speed than does the low frequency noise. This high frequency peak is cavitation noise, mainly from the propellers, so high frequency self noise can be reduced considerably by excluding noise arriving from the direction of the propellers. This causes a blind spot for sonar detection, but improves performance in other direc-

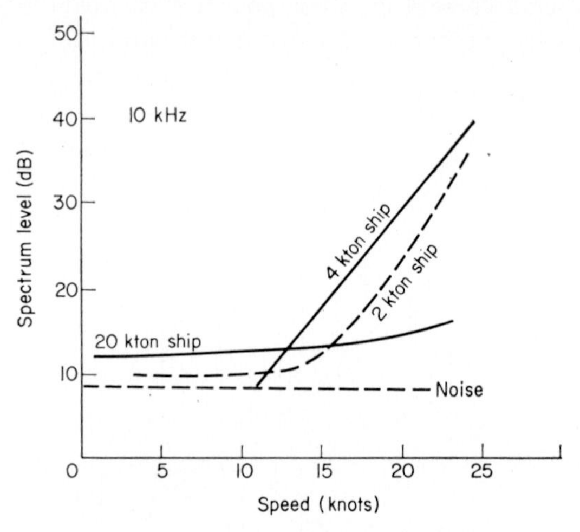

Fig. 4.20 Ambient noise in sonar dome

tions. The methods of excluding the cavitation noise are two fold. Firstly the sonar beam can be made directional and secondly attenuating material can be placed between the propellers and the sonar transducer.

Fig. 4.20 shows a cross-section of Fig. 4.19 taken at 10 kHz. To it has been added representative shaped curves for other types and sizes of ship. A very general summary of the main causes of self noise may be subdivided thus:

(a) below cruising speed—auxiliary machinery at all frequencies,
(b) above cruising speed—machinery noise from main engines up to 2 kHz
—cavitation noise above 2 kHz.

4.4.4 *Ambient noise*

This is the natural noise in the ocean and is of importance in fixed hydrophone arrays used for passive detection, where there is no reverberation and no self noise. It is made up of:

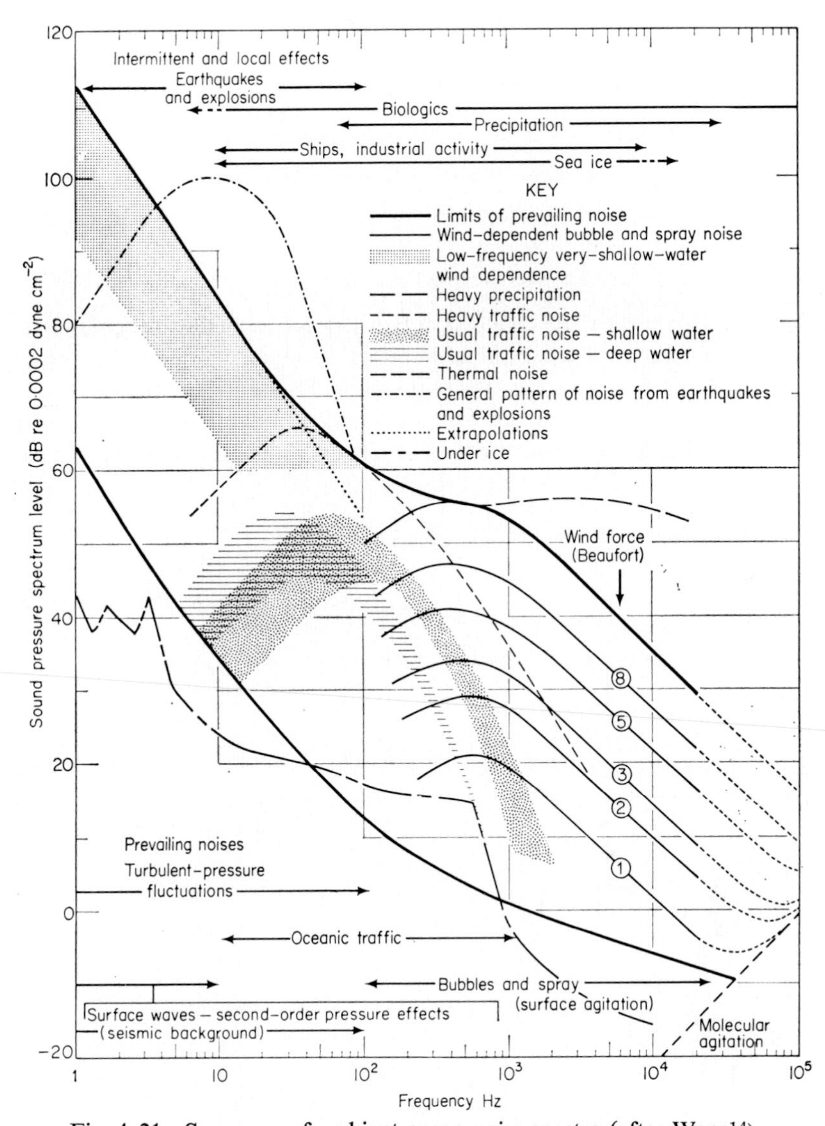

Fig. 4.21 Summary of ambient ocean noise spectra (after Wenz[14])

(a) Physical effects—surface motion, rain, earth movements, etc.
(b) Biological noise—fish, etc.
(c) Man-made noise—shipping and industry.

Wenz[14] identifies 4 overlapping regions:

(a) above 30 kHz—molecular thermal agitation $p \propto f$

$$L = -115 + 20 \log f_{\text{kHz}} \text{ decibels re 1 microbar at 15 °C} \tag{4.32}$$

(b) 100–30,000 Hz—wind induced surface agitation,
(c) 10–500 Hz—shipping,
(d) below 10 Hz—turbulence effects.

The lowest ambient noise levels have been measured below ice, emphasizing the importance of wind induced surface agitation.

Lomask and Frassetto used the bathyscaph Trieste to measure the variation of ambient noise up to 300 Hz as a function of depth in deep water. In sea state 2 (calm) they found the level decreased with depth from the surface to 3000 metres as follows:

(a) at 68 Hz by 16 dB,
(b) at 10 Hz by 6 dB.

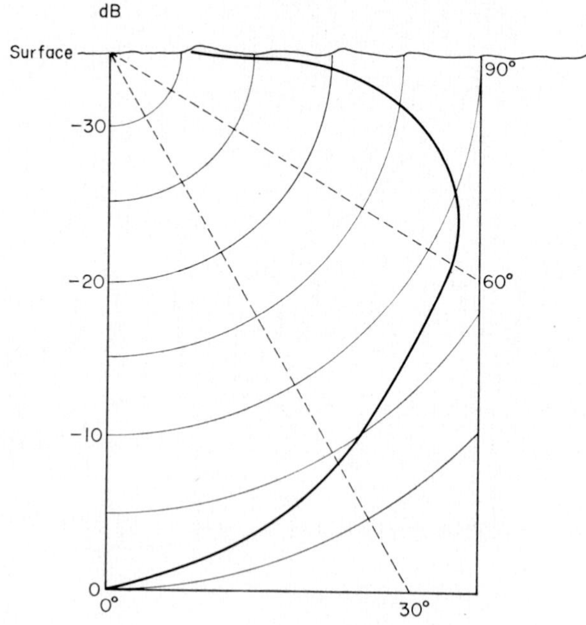

Fig. 4.22 Directivity of the ocean surface as a source (after Becken)

In heavy rain, a shallow hydrophone records about 15–25 dB increase in noise level over the frequency range 100–30,000 Hz. There is one report of a 35 dB increase of noise level on a shallow hydrophone at 10 kHz when a half inch of rain fell in $1\frac{1}{2}$ hours.

(i) *Directionality* Until recently[15–17], ambient noise was mistakenly assumed to be isotropic. The noise from whitecaps and bubbles is directional in both azimuth and tilt. In azimuth there is a maximum of noise in the direction of the troughs of the prevailing swell and a minimum normal to the swell waves. In the vertical plane the surface can be considered as a source with the directivity pattern shown in Fig. 4.22.

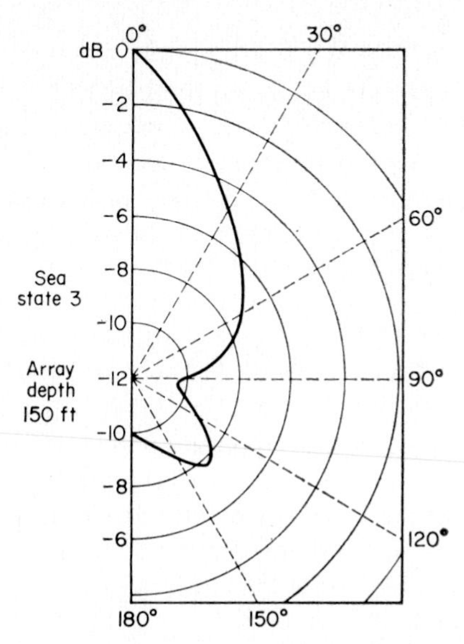

Fig. 4.23 Typical directionality of sea noise at a depth of 150 ft (after Becken)

There is usually a significant contribution to the measured noise from surface generated noise reflected back from the bottom of the ocean. The amount scattered increases as the grazing angle becomes smaller, so if we measure ambient noise at a depth of 150 ft we get the vertical directional pattern shown in Fig. 4.23.

Here the null at the waist represents the low horizontal strength of the surface source and the minimum downwards indicates the poor scattering of the ocean floor at normal incidence. The lobe at about 140° indicates that here the combination of surface source directionality and improved scattering at this angle more than makes up for the extra range of propagation.

4.5 Conclusion

It has not been possible to discuss every aspect of acoustic exploration, so the terms which feature in the sonar equations have been considered. These alone are so complex that only a superficial treatment could be undertaken in the space available. The propagation loss (P_2), which we did not treat, is a large discipline on its own with a vast specialist literature. Signal processing, which has aspects common to communications and radar is probably even bigger. Modern advances in exploration depend on bringing these disciplines together and taking due account of all the complexities which have been stressed in this chapter.

Acknowledgements

Figs. 4.17 and 4.18 were kindly supplied by Dr. J. W. English of the Ship Division of the National Physical Laboratory, Feltham. Figs. 4.5, 4.9, 4.10 and 4.11 were made specially for this chapter on the Pinger Tracking Equipment at the National Institute of Oceanography, Wormley. Figs. 4.19 and 4.20 are based on measurements made by the Royal Australian Navy Research Laboratory, Sydney.

Appendix—glossary of terms

S_A — SOURCE STRENGTH: equivalent intensity in the bandwidth of the receiver at unit distance from the source in the direction of the target based on the propagation laws pertaining to the whole propagation path (i.e., not necessarily the intensity which would be measured at unit distance from the source). The most common unit used for distance is 1 yard.

S_P — SOURCE STRENGTH of the target in the direction of the observer (passive detection).

N — TOTAL MASKING BACKGROUND NOISE: equivalent noise in the bandwidth of the receiving equipment referred to the input to the receiver in the same units as the source strength.

P_2 — TWO-WAY PROPAGATION LOSS: includes spreading, interference and absorption. Note that the loss can be different in each direction.

P_1 — ONE-WAY PROPAGATION LOSS for passive detection.

T — TARGET STRENGTH: source strength of a target per unit incident intensity = 10 log (ratio of the equivalent reflected intensity to incident intensity) at unit distance from the target. (The most common unit distance is 1 yard.)

S_x — EXCESS ECHO LEVEL at the receiver.

R — RECOGNITION DIFFERENTIAL: $\Delta - G_\Delta - S_x$.

$\Delta - G_\Delta$ DETECTION THRESHOLD. The input signal to noise ratio at the receiver corresponding to some stated output signal to noise ratio which determines the false alarm rate for some given detection probability.

G_Δ PROCESSING GAIN of the detection system for an output signal to noise ratio of Δ.

References

1 Urick, R. T., 'Prediction Methods for Sonar Systems', *J. Br. Instn. Radio Engrs*, **25**, no. 6, 501 (1963).

2 Tucker, D. G., In *Underwater Acoustics* (V. M. Albers, ed.), p. 29, Plenum, New York, 1961.

3 Stocklin, P. L., In *Underwater Acoustics* (V. M. Albers, ed.), p. 339, Plenum, New York, 1961.

4 Siebert, W. McC., 'A Radar Detection Philosophy', *I.R.E. Trans. Inf. Theory*, PG IT-2-1955–56, 204 (1956).

5 Stewart, J. L., and E. C. Westerfield, 'A Theory of Active Sonar Detection', *Proc. Inst. Radio Engrs* **47**, no. 5, 872 (1959).

6 Federici, M., 'On the Improvement of Detection and Precision Capabilities of Sonar Systems', *J. Br. Instn. Radio Engrs*, **25**, no. 6, 535 (1963).

7 Brekhovskikh, L. M., 'Possible Role of Acoustics in the Exploring of the Ocean', in 'Reports of General Conferences', *5th International Congress on Acoustics*, Vol II (D. E. Commino, ed.), Liege, Belgium, 1965.

8 Andreeva, I. B., 'Acoustical Characteristics of Sonic Scattering Layers in the Ocean', *5th International Congress on Acoustics*, Vol. I, paper E 68, Liege, Belgium, 1965.

9 Barham, E. G., 'Deep Scattering Layer Migration and Composition: Observations from a Diving Saucer', *Science*, **151**, 1403 (1966).

10 Ackroyd, J. O., 'The Detection of Sonar Echoes in Reverberation and Noise', *J. Br. Instn. Radio Engrs*, **25**, no. 2, 119 (1963).

11 Heller, S. R., Jr., 'Hydroelasticity', in *Advances in Hydroscience* (Ven Te Chow, ed.), Vol. 1, p. 94, Academic, New York and London, 1964.

12 Ffowcs Williams, J. E., In *Underwater Acoustics* (V. M. Albers, ed.), Vol. 2, p. 89, Plenum, New York, 1967.

13 Skudrzyk, E. J., and G. P. Haddle, In *Underwater Acoustics* (V. M. Albers, ed.), p. 255, Plenum, New York, 1961.

14 Wenz, G. M., 'Acoustic Ambient noise in the Ocean: Spectra and Sources, *J. Acoust. Soc. Am.*, **34**, no. 12, 1936 (1962).

15 Becken, B. A., 'Sonar', in *Advances in Hydroscience* (Ven Te Chow, ed.), Vol. 1, p. 1, Academic, New York and London, 1964.

16 Albers, V. M., *Underwater Acoustics Handbook* 11, The Pennsylvania State University (1965).

17 Rudnick, P., V. C. Anderson and Cdr. B. A. Becken, 'Directional Distribution of Ambient Sea Noise', *J. Br. Instn. Radio Engrs*, **25**, no. 5, 441 (1963).

5

Acoustic Echoes from Targets under Water

R. W. G. Haslett

Kelvin Hughes, a Division of Smiths' Industries Ltd., Hainault, Essex

---

5.1 Introduction

Although the reception of echoes from reflecting or scattering targets is well known in radar, and also in ultrasonic flaw-detection and in underwater acoustic echo-ranging, the prediction of the strength and character of the echoes is a matter of some complexity except when the target has a simple shape. This chapter, relating to acoustic echoes from targets under water, is largely based on the work of the author. (Further details can be obtained from the published papers listed in the references.)

Many echo-ranging equipments in the field of underwater acoustics transmit single pulses at a fixed frequency and these give rise to echoes which are complicated enough. However, the character of the echo is even more intricate when other types of transmission are used, for example, a frequency-modulated pulse in which the frequency steadily varies according to a special law or a train of pulses at differing frequencies. Pulses of random noise contained in a certain bandwidth may also be used or, alternatively, pseudo random-noise pulses which repeat the same pattern on each transmission.

In the case of a complex target which is large compared with the wavelength, at a fixed frequency interference occurs between the echoes from various parts of the target which cause the received signal to vary rapidly in amplitude when the target is rotated. This effect is increased with frequency-modulated pulses when the amplitude of the echo passes through all possible maxima and minima in the course of the frequency sweep. In contrast, when using pulses of noise, the signals are averaged and the interference effects disappear. Consequently, orientation of the target produces only a gradual change in the echo level.

It is proposed to consider echoes obtained from single pulses transmitted at fixed frequency, representing the simplest case. When the pulse is long (for example, more than ten wavelengths), the conditions at the centre of the pulse may be taken to be substantially as if the waves were continuous. Initially, the echoes from the targets will be assumed to be distinct from those received from the sea bed or from the water surface.

5.2 Basic principles of operation of a typical echo-ranger

Many types of recording echo-ranging equipments embody the features shown diagrammatically in Fig. 5.1[1]. The interval between pulses is determined by the time of rotation of a contact drum, synchronized with the movement of a pen at constant speed across electro-sensitive recording paper. The path of the pen is either a wide arc of a circle or a straight line. Each time the pen comes on to the paper, the transmission contact closes and a pulse of high-frequency alternating current is applied to the transmitting

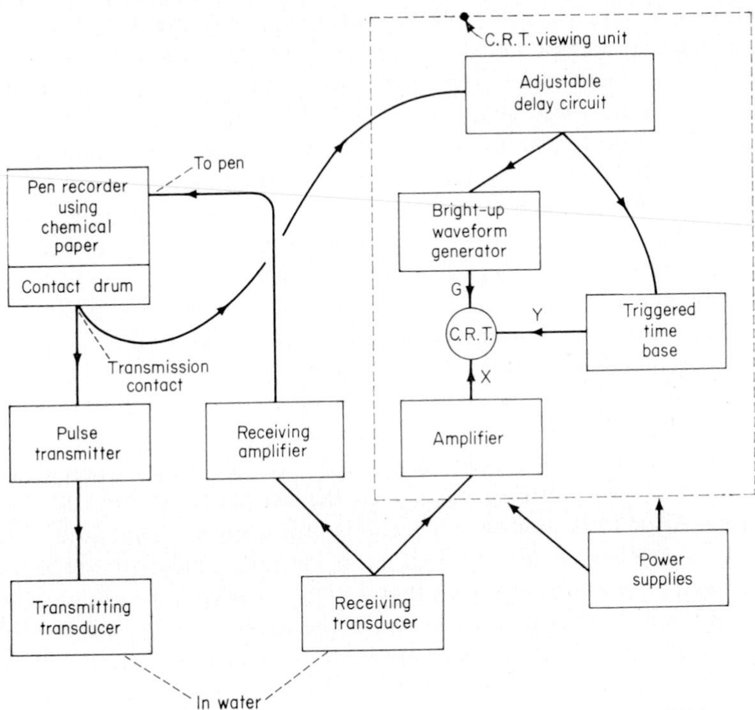

Fig. 5.1 Block diagram of a typical echo-sounder (with the addition of a cathode-ray viewing unit)[1]

transducer, mounted underwater. The frequency is usually between 5 and 300 kHz and the pulse length from 100 μs to 100 ms. Sometimes a form of time-varied gain is used on the receiver whereby the sensitivity is reduced at short range. In this case, the controlling circuit is triggered by the transmission contact.

Echoes from discontinuities in the acoustic impedance of the water (resulting from the natural environment or targets) impinge on the receiving transducer and are amplified. These signals cause an electric current to flow from the pen through the paper to give an intensity-modulated record on which the traverse of the pen represents a range scale. To this equipment a cathode-ray tube display can be added. If the latter is triggered from the transmission contact through an adjustable delay, a certain part of the range scale can be expanded for more detailed examination of echoes.

5.3 Types of target met in various applications

Equipment based on the above principles are used in many fields. Chief among these is in the determination of depth when navigating ships at sea and in hydrographic survey of the oceans or of in-shore waters. In these cases, the acoustic beam is vertical and the echoes are received from the sea bed. Their strengths and characters depend on the roughness of the sea bed and its acoustic reflectivity.

Another major use is for detecting fish, either close to the sea bed or in mid-water. In sea-bed trawling, the small echoes from individual fish are often detected separately from one another in depths of 200 fathoms and there is the added difficulty of distinguishing them from the sea-bed echo which is very much larger. This difficulty does not arise in mid-water trawling, in which large shoals of fish are met and conglomerate echoes of high amplitude result from the superposition of contributions reflected from large numbers of fish.

In the Asdic type of equipment, the axis of the acoustic beam is near the horizontal, and can be rotated to search in various directions from the ship and adjusted in tilt. As a result, targets are detected against a general background of sound scattered back from the sea bed and from the water surface. Apart from military applications, this method is used in detecting shoals of fish and whales, as well as in the examination of wrecks and underwater constructions in civil engineering. Geological surveys use a fan-shaped beam (narrow in the horizontal plane and wide in the vertical plane) fixed to the side of the ship, with the axis of the beam tilted slightly downwards, so that much of the projected sound meets the sea bed at glancing angles. If the ship moves forward at a steady speed along a straight track, a form of acoustic 'map' of the sea bed is produced on the recorder[2]. The received echoes are affected largely by the topography of the sea bed and its 'roughness' in comparison with the wavelength. Interpretation of the record is aided by examination of the acoustic 'shadows' which occur behind regions which stand proud of the sea bed.

The closest observation of all these echoes can be made on a cathode-ray display in which the vertical deflection is proportional to amplitude of echo and the horizontal time-base represents range. In this way, the individual cycles of the wave may be seen. Chemical recorders and other intensity-modulated displays give less detail but give a wider appreciation of a larger area and of the shapes of correlated echoes. The latter applies to plan-position displays such as those produced by sector scanning in which a narrow acoustic beam is rapidly swept over an angular sector. Another sophisticated piece of equipment for viewing underwater targets is an experimental closed-circuit acoustic television system which produces an acoustic 'picture' of the target on a monitor screen[3]. An acoustic 'image' of the target is transformed to a television picture by using a special form of acoustic to electronic converter tube rather like a television camera tube. Here again, intensity modulation is used in the display.

Although there is a wide diversity in the echoes received from various targets, a number of common basic principles can be used to determine the strengths and characters of the echoes and their dependence on target aspect. The results for those targets which have been studied in detail, can be applied to many other types of targets.

5.4 Propagation of acoustic waves under water—sonar equation

The propagation of the acoustic waves to the target and back can be analysed mathematically[1] assuming ideal conditions, namely:

(a) an infinite homogeneous medium
(b) good acoustic contact between the transducers and the water
(c) the distances from the transducers to the target are large compared with their Fresnel zones
(d) the use of a small target compared with the width of the beam
(e) the spreading of the energy follows an inverse-square law from the transmitting transducer to the target
(f) the spreading of the energy follows an inverse-square law from the target to the receiving transducer
(g) the effects of refraction by thermoclines, pressure and salinity gradients, are small, and
(h) the pulse is many cycles long (at least ten) so that the centre of the pulse may be considered to be continuous wave.

The treatment can then be extended to take account of the transducer parameters. Thus, in the centre of an echo-pulse from a target at angle θ off the axis of the beam, the resultant power (W'_θ) dissipated instantaneously in the receiving transducer load, is related to the electrical power (W) accepted by the transmitting transducer, by the equation

$$W'_\theta = \frac{W \cdot 10^{-0 \cdot 2\alpha r} \sigma \eta_{MA} \eta_{EM} \eta'_{MA} \eta'_{EM} \lambda^2 D_\theta^2 D_\theta'^2}{64\pi^3 r^4 (\eta_D)_f (\eta'_D)_f} \qquad (5.1)$$

where α is the acoustic attenuation in the water (dB cm^{-1})
σ is the acoustic back-scattering cross-section of the target (cm^2)†
λ is the wavelength in water (cm)
r is the range (cm)
η_{MA} is the mechano-acoustic efficiency
η_{EM} is the electro-mechanical efficiency
$(\eta_D)_f$ is the directivity factor
} of the transmitting transducer
D_θ is the relative pressure directivity factor of the transmitting transducer at angle θ off the axis (compared with the axial ray)

and other similar terms (dashed) refer to the receiving transducer.‡ $(\eta_D)_f$ concerns the pressure on the axis of the beam and D_θ relates to the polar diagram. The mechano-acoustic and electro-mechanical efficiencies of the transducers can be determined by bridge measurements.

The acoustic attenuation in fresh water and in sea water rises steeply with frequency, as seen in Fig. 5.2[4].

In practice, this formula serves as a guide to the echo to be expected from a target of a given cross-section, but is more accurate in vertical echo-sounding in which the sound is propagated normally to the surfaces containing points having the same acoustic velocity (since the velocity gradients are usually vertical), rather than in horizontal ranging which is more prone to bending of the beam.

Knowing the following factors:

(a) the directivity patterns of the transducers
(b) the acoustic attenuation of the water
(c) the angle (θ) between the line joining the target to the transducer and the axis of the beam, and
(d) the law of time-varied gain

† σ is defined as the plane area intercepting an amount of *energy*, which, if it were scattered from the target uniformly in all directions, would produce an echo equal to that observed. (This area is placed at the same position as the target, perpendicular to the direction of propagation of the incident waves.)

‡ Sometimes the various terms in the sonar equation are expressed in dB relative to specified quantities[5], for example the back-scattering cross-section is given as a *target strength* in dB relative to the area of cross-section of a sphere of radius 2 yd (viz: $1{\cdot}05 \times 10^5$ cm^2).

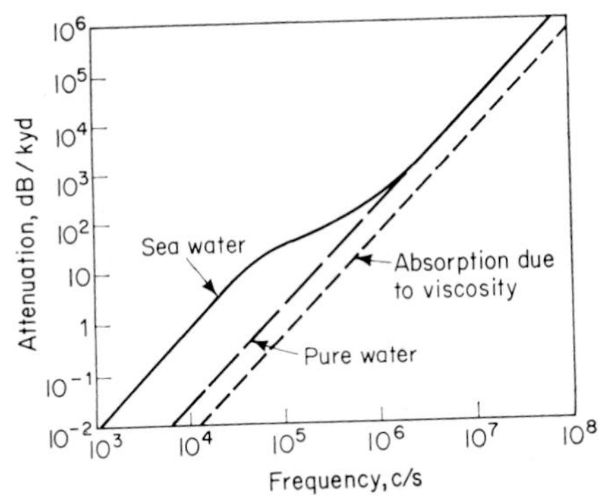

Fig. 5.2 Attenuation of acoustic waves in water; its dependence on frequency. Graph for sea water after Reference 5: graph for pure water is due to Schulkin and Marsh[6]†

it is possible to draw diagrams representing a series of 'concentric' calibrated fields of detection of a single target beneath a ship, consisting of pear-shaped surfaces. Each surface is the locus of points at which a target of a given cross-section gives a uniform received echo (at a known gain-setting).

Confirmation of one of these zones of detection may be seen in the work of Sünd[7] (among others) who lowered a compact target consisting of twenty air-filled glass floats, to map the actual acoustic beam (Fig. 5.3). There is close agreement between his results and the radiation diagram (calculated by the present author) after correction for time-varied gain.

Measurement of the echoes received from targets at sea is much more difficult than in a water tank in the laboratory, due to the changing conditions and the motions of ship, target and water. Thus, much more accurate measurements of the acoustic cross-sections of targets have been made in the laboratory, especially in regard to orientation of the target under carefully-controlled conditions.

5.5 Scale-model measurements

A further facility is the use of a scale-model technique[8] in which the dimensions of the target are reduced by a convenient scaling factor and the frequency of operation is raised by the same factor. For example, a frequency of 1·5 MHz can be used in the model, which gives a scaling factor of 50/1 as compared with a full-size frequency of 30 kHz. When there is such a large reduction in the space required, a water tank can be used.

As compared with measurements at sea, this method has the great merits

† Being reproduced from the original papers, some figures in this chapter refer to 'c/s', 'kc/s' and 'Mc/s' and should be read as Hz, kHz and MHz respectively.

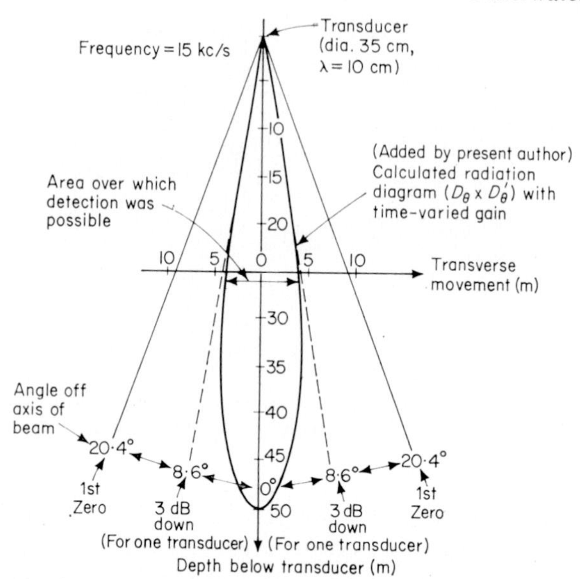

Fig. 5.3 The zone of detection beneath the ship, observed by Sünd (correlated with calculation)[1, 7]

of stable conditions for measurement, of accurate adjustment of targets and repeatability of results.

The following details require consideration when deciding the scaling factor:

(a) size of water tank
(b) ability to make the model targets of sufficient accuracy
(c) attainment of an adequate target/reverberation and target/support echo-ratios
(d) beam angles of the transducers
(e) overall bandwidth of the system, and
(f) the minimization of the errors due to incorrect scaling of acoustic absorption in water (see Fig. 5.2).

The last effect requires that the target is short or that the attenuation along the length of the target is small or that only very short pulses are used with time-varied gain to correct the amplitude of the envelope of the echo. This error can be considerable, especially for large targets and long pulses.

Fig. 5.4 is a block diagram of a scale model in which four trigger pulses were derived from the mains supply. The first (1) controlled the pulse transmitter which applied an approximately rectangular pulse at 1·5 MHz to the transmitting transducer. After a delay, three further trigger pulses

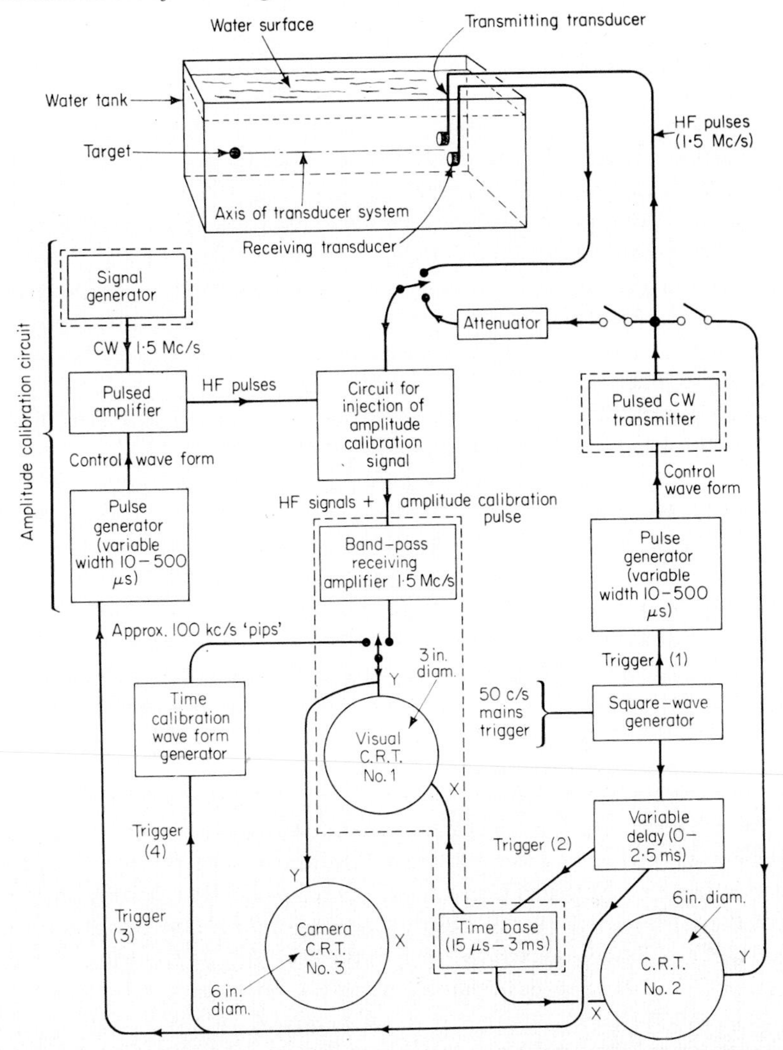

Fig. 5.4 Block diagram of apparatus for scale-model measurements[8]. (The portions enclosed by the dotted lines were manufactured instruments)

were obtained. The second pulse (2) triggered the time base so that the received echoes could be amplified as HF envelopes on the cathode-ray tube No. 1. The variable delay allowed expansion of any portion of the echoes for detailed examination.

The remainder of the apparatus was used for calibration of frequency, amplitude and time, the standard of frequency being the signal generator.

For amplitude calibration, the incoming echo was compared with a synthetic HF pulse of known amplitude, which was triggered by waveform (3) and appeared at the start of the time base. The amplitude of this pulse was related directly to the reading of the attenuator on the signal generator. For routine time-calibration, marker 'pips', spaced at intervals of approximately 10 μs were also locked to the time base (trigger 4). For more accurate measurement of time, a signal generator was applied to the time calibration waveform generator. The amplitude, shape and length of the electrical pulse across the transmitting transducer were observed on cathode-ray tube No. 2. By using an attenuator between the transmitter and the receiver, the transmitted pulse was also visible on cathode-ray tube No. 1. A separate tube (No. 3) was used for photography.

The two transducers, each adjustable in angle, were fixed at one end of the tank (6 ft long) with their beams pointing at the target.

After much experiment, a rigid target support was evolved using $\frac{1}{4}$-inch diameter steel rod which was polished and bent so that no part of the support in the acoustic beam appeared in broadside aspect. The target was mounted on the point of a needle attached to the lower end of this support. This arrangement allowed accurate rotation of the target about a vertical axis and a high target/support echo-ratio was obtained over a wide angle of rotation.

The scale-model technique is adaptable, for example, a frequency of 625 kHz corresponds to a scaling factor of 21/1 as compared with 30 kHz, whilst 360 kHz gives a scaling factor of 12/1.

5.6 Acoustic cross-sections of small targets, amenable to calculation

The back scattering cross-sections of various simple bodies in the geometrical region (i.e., target much larger than the wavelength) may be calculated in terms of the dimensions and of the acoustic reflectivity, by analogy with the reflection of electro-magnetic waves[4, 9, 10] (Table 5.1). $\mathcal{R}$ is the acoustic amplitude reflectivity (per cent) for the body material (see Section 5.7). These values relate to plane incident waves at a single frequency and to targets which are small compared with the width of the acoustic beam. The amplitude (or voltage) of the received echo is, of course, proportional to the square root of the cross-section.

The list of body shapes for which the acoustic cross-section is readily amenable to calculation is restricted to those seen in this table, together with a few other simple shapes. Each of these formulae is for a body made from a material which has an acoustic impedance greater than that of water and is for the front surface of the body, only, ignoring interference effects due to echo contributions from the rear surface.

This cross-section relates to the 'high-light' in broadside aspect which will

Table 5.1 *Back-scattering cross-sections of the front surfaces of simple bodies in the geometrical region*[4]

Body	Aspect	Cross section (σ)	Nomenclature
Entirely convex surface	Any	$\pi a_1 a_2 \mathscr{R}^2$	a_1 and a_2 are the two principal radii of curvature and $\mathscr{R}$ is the amplitude reflection coefficient
Sphere	Any	$\pi a^2 \mathscr{R}^2$	a is the radius
Ellipsoid	Broadside	$\dfrac{\pi L^2 H^2 \mathscr{R}^2}{4B^2}$	L is overall length H is overall height and B is overall breadth
	End-on	$\dfrac{\pi H^2 B^2 \mathscr{R}^2}{4L^2}$	
Short circular cylinder	Axis inclined	$\dfrac{a'\lambda}{2\pi} \dfrac{\cos\phi \sin^2(kl\sin\phi)}{\sin^2\phi} \mathscr{R}^2$	ϕ is the angular deviation from broadside aspect l is the length and a' the radius of the cylinder $k=\dfrac{2\pi}{\lambda}$
	Broadside ($\phi=0$)	$ka'l^2\mathscr{R}^2=\dfrac{2\pi a'l^2\mathscr{R}^2}{\lambda}$	
Large plane rectangular plate	Inclined	$\dfrac{4\pi A^2}{\lambda^2}\left[\dfrac{\sin(kb\sin\phi)}{kb\sin\phi}\right]^2 \cos^2\phi\mathscr{R}^2$	One side is perpendicular to the incident direction, the other side (of length b) makes an angle ϕ with the incident direction A is the area of the plate
	Broadside ($\phi=0$)	$\dfrac{4\pi A^2\mathscr{R}^2}{\lambda^2}$	

run across the body when it is rotated about a vertical axis, as occurs with specular bodies using light waves. However, since the wavelength is comparatively large and the energy is coherent, a polar diagram having side-lobes is observed, for example, when a short circular cylinder or a plane rectangular plate is inclined at angle ϕ, as given in the third column of Table 5.1.

Thus, the target may be considered to have a polar diagram of back scattering in three dimensions, which becomes very complicated when the target is many wavelengths long. The distribution of energy is broadly related to that around a radiating transducer of similar shape, except that there are twice as many lobes in the polar diagram for the target as for the transducer, over the same range of angles. σ varies with the size and shape of the target and with its orientation. In practice, since the incident waves usually depart from ideal uniform plane waves, there is some variation in σ with range, especially at short ranges.

5.7 Acoustic reflectivities of various materials

The values of the acoustic amplitude reflectivities ($\mathscr{R}$) of various substances are seen in Table 5.2[4, 11]. These are for normal incidence on an ideal plane interface between water and the material. The reflectivity is

Table 5.2 *Acoustic reflectivities of various substances in water*[4]

Substance	Acoustic impedance, c.g.s. units	$\mathscr{R}$ in tap water, %	$\mathscr{R}$ in salt water, %
Air	41	−100	−100†
Steel	$4 \cdot 7 \times 10^6$	94	94
Brass	$3 \cdot 8 \times 10^6$	92	92
Aluminium	$1 \cdot 7 \times 10^6$	84	83
Granite	$1 \cdot 6 \times 10^6$	83	82
Quartz	$1 \cdot 5 \times 10^6$	82	81
Clay	$7 \cdot 7 \times 10^5$	68	67
Sandstone	$7 \cdot 6 \times 10^5$ (approx.)	68	66
Perspex	$3 \cdot 06 \times 10^5$	35	33
Wet fish bone	$2 \cdot 5 \times 10^5$	26	24
Rubber (pencil eraser)	$1 \cdot 81 \times 10^5$	11	8
Wet fish flesh	$1 \cdot 60 \times 10^5$	4·6	1·9
ρc rubber	$1 \cdot 55 \times 10^5$	3·0	0·32
Tap water at 15°C	$1 \cdot 46 \times 10^5$	—	—
Sea water (salinity: 35 parts per 1000, 15°C)	$1 \cdot 54 \times 10^5$	—	—

(At 1·5 MHz)

† The negative sign corresponds to a change of phase.

related to the acoustic impedances of the substance (Z_2) and of the water (Z_1)

$$\mathscr{R}=\frac{Z_2-Z_1}{Z_2+Z_1}$$

where $Z=\rho c$, ρ being the density and c the acoustic velocity in the material.

For comparison, the acoustic impedances of tap water and of sea water are also given.

Generally speaking, materials whose densities differ greatly from that of water are good reflectors of sound, for example, the heavy metals, rocks and air bubbles, whilst soft materials with densities close to that of water are poor reflectors. The presence of even a small amount of air within a body can increase the strength of the echo, since air is such a good reflector of sound ($\mathscr{R}=-100$ per cent).

When the reflectivity of the front surface of the body is less than 100 per cent, some of the sound penetrates into the material and is reflected by the rear surface. As a result, interference between these two contributions to the echo can occur and the effect can be extremely complex (for example, see Section 5.11.2 regarding a cylinder, translucent to sound).

5.8 Standard targets

Although in scale-model and in full-size measurements it is possible to measure all the parameters of the system and so determine the back-scattering acoustic cross-section of the target from Eqn. 5.1 (Section 5.4) assuming ideal propagation, it is more convenient to use a standard target (which is substituted for the actual target) to calibrate the system. Moreover, this eliminates any long-term effects due to anomalous propagation.

In order to reduce the error due to the interference between front and rear surfaces of the target (mentioned in the last section) to a negligible proportion of the returned echo, an almost rigid material must be used, alternatively an air-backed surface (see Table 5.2).

An air-filled sphere should theoretically serve as a good standard target, having the added advantage of a cross-section which is independent of frequency. In practice, the sphere must be made thin to minimize the interference effects due to the shell.

The use of an air-filled metal sphere at sea as a standard target is demonstrated in Fig. 5.5[12] in which a comparison of received signals is made with those from fish lowered on a wire to various depths. The accuracy of measurement is, in any case, not as high as in a scale model. It may be observed that the general decline in echo amplitude with depth follows an inverse-square law, plus a further reduction for acoustic attenuation, as predicted by Eqn. 5.1 (since the acoustic cross-section is proportional to the square of signal amplitude). In this case, the target is small compared with the width of the beam. The mean acoustic cross-section of the fish approximates to that of the sphere, namely $\pi a^2 = 314\ \text{cm}^2$.

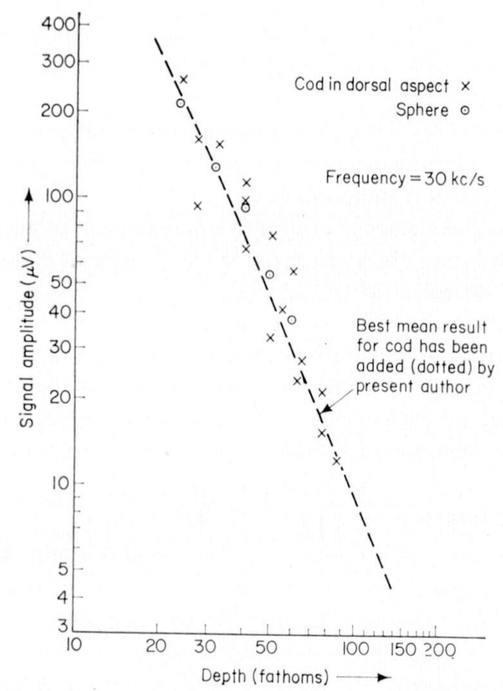

Fig. 5.5 Harden-Jones's results for Cod (of mean length 85 cm) and a 20-cm diameter air-filled spherical metal float suspended on a wire[1]

A solid metal sphere has been found to be unreliable (possibly due to inhomogeneities) but a solid metal cylinder has proved to be accurate as a standard target provided it is carefully orientated to lie precisely in broadside aspect. In this case, the calculated acoustic cross-section in the geometrical region (from Table 5.1)

$$\sigma = 2\pi a' l^2/\lambda \qquad (5.2)$$

and is seen to be proportional to frequency.

5.9 Acoustic cross-sections of large targets

When no longer small compared with the width of the beam, the target may be treated as non-uniformly irradiated and a correction might be applied for this. Also, only those parts of the target which are separated by an echo distance less than the pulse length in water can combine to give the conglomerate echo at any instant.

In the extreme case of a very extensive target, the actual part which contributes to the observed cross-section at any instant is determined by the shape and width of the beam and by the length of the pulse. This volume is known as the instantaneous 'pulse packet' and increases in area with range.

The effect is illustrated in Fig. 5.6 in which the echo levels returned by a fish and by the sea bed are measured on a deep-sea trawler[4, 13], with respect

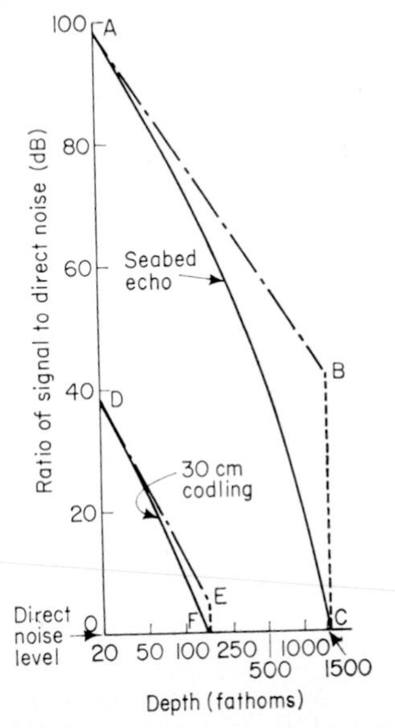

Fig. 5.6 Relative signal levels received on a typical echo-sounder at 30 kHz in good weather at sea (after Hopkin)[13]. The lines AB and DE are obtained after the effects of attenuation in the water have been eliminated. The direct noise is noise from the propeller which has travelled to the transducer by the shortest path[4]

to a convenient fixed level (corresponding to the minimum noise-level observed when towing the trawl, known as the 'direct noise-level'). The pulse length was 0·5 ms and the mean beam width 21°. As in Fig. 5.5, the signal level for the 30-cm codling follows Eqn. 5.1 with depth (r). After subtracting the ordinate FE (which equals the attenuation from 20 to 125 fathoms and back), the straight line DE shows the required inverse fourth-power law of variation with depth.

In contrast, the sea-bed echo follows an inverse-cube law (line AB, after subtracting the absorption from 20 to 1500 fathoms and back, represented by the ordinate CB). In this case, it is normal to use Eqn. 5.1 still, but to postulate an acoustic cross-section proportional to range. The sea bed is rough and acts rather like a diffuse reflector. The area of the sea bed within the pulse packet is, in this case, restricted by the pulse length and is proportional to depth.

Similar results have been obtained using three different targets in scale-model measurements (Fig. 5.7)[4]. The upper graph is for a very small target which consisted of a plane cork disc of diameter 1 mm ($\mathscr{R} = -100$ per cent), which was placed perpendicular to the axis of the beam at each range. Its

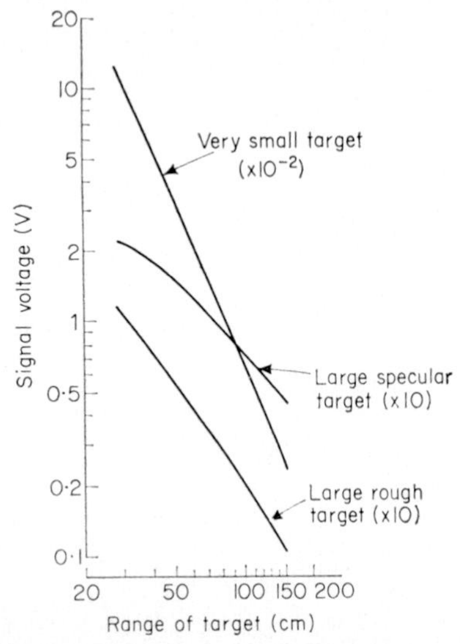

Fig. 5.7 Scale-model measurements. Variations of echo amplitudes with range for various targets. N.B. The ordinate must be multiplied by the constant shown in brackets for each graph[4]

scale acoustic cross-section (from Table 5.1) was about 8×10^{-2} cm^2 which corresponds to a fish of length about 40 cm at 30 kHz. (Acoustic cross-section, being an area, is scaled according to the square of the scaling factor.) In Fig. 5.7, this target is seen to obey the inverse fourth-power law, plus the attenuation of water. (The effect of attenuation is relatively small, raising the graphs at a range of 150 cm by a factor of 1·21/1, only.)

The 'large specular target' was a plane sheet of aluminium 30 × 30 cm and 0·161 cm thick. As 0·161 cm equals 0·43 of the wavelength in aluminium, this target will have a total reflectivity of 90 per cent (see Fig. 5.18 and Section 5.12.2). A reflected 'image' of the transmitting transducer will appear in the plate so that the echo intensity would fall off inversely as the square of range (after the attenuation of the water is allowed for). This is indeed found to be so.

The 'large rough target' consisted of a similar plate covered by a layer of brass turnings. The main dimensions of these particles (measured under a travelling microscope) varied between 0·12 mm and 2·5 mm (0·12 λ and 2·5 λ at 1·5 MHz in water). These corresponded to stones of lengths from 0·6 to 12·5 cm at 30 kHz on the sea bed. The pulse length was 25 μs, equivalent to 1·25 ms at 30 kHz and the beam angle was 9° between zeros. The signal power received from this target was seen to vary in inverse proportion to the cube of range (after allowance for attenuation), just as had been discovered previously at sea (Fig. 5.6).

The actual back-scattering cross-sections in Figs. 5.6 and 5.7 can also be correlated when due allowance is made for changes of beam width and pulse length and the lower reflectivity of the sea bed as compared with brass.

It must be inferred therefore that the sea beds met whilst trawling are diffuse reflectors. (A confirmation of this is seen in the observed sea-bed echo which is found to be many times longer than the transmitted pulse.) In fact, the sea bed is known to consist of shingle, corrugated sand and clay.

5.10 Cross section of a sphere

It is to be noted that in the case of an entirely convex surface (such as a sphere or ellipsoid), the cross section in the geometrical region is independent of the frequency.

Kerr[10] has indicated how the back-scattering cross-section of a perfectly-conducting sphere (Fig. 5.8) varies with radius of sphere a in the electromagnetic case. This corresponds to a rigid sphere in acoustics. In the Rayleigh scattering region below $a/\lambda = 0{\cdot}16$, the back-scattered energy declines sharply, being asymptotic to the inclined broken line (which represents a fourth-power law).

Above this value, there is a region of regular fluctuation which declines in amplitude, falling below 10 per cent error from the mean above $a/\lambda = 1{\cdot}3$. The curve is asymptotic to the horizontal broken line, representing a constant value of $\sigma = \pi a^2$ in the geometrical region, independent of frequency.

5.11 Cross section of a short cylinder

5.11.1 Cylinder opaque to sound

As seen in Table 5.1, the back-scattering cross-section of a perfectly-conducting short cylinder in broadside aspect was examined by Mentzer[9],

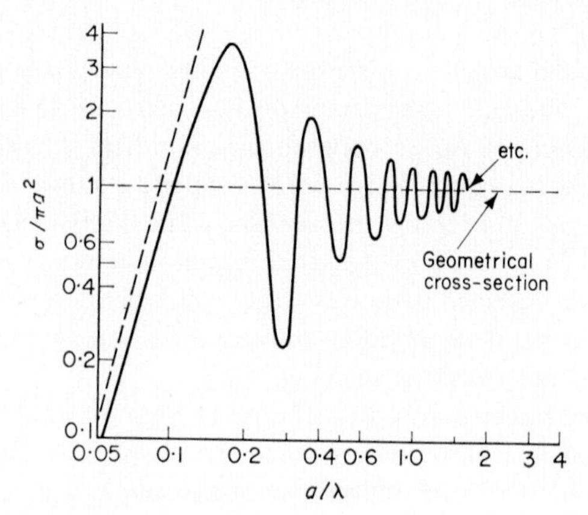

Fig. 5.8 Ratio of back-scattering cross-section (σ) to geometrical cross-section (πa^2) in the radar case of a perfectly reflecting sphere. a=radius of sphere and λ=incident wavelength. The inclined broken line represents Rayleigh's law, $\sigma/\pi a^2 = 1{\cdot}403(a/\lambda)^4 \times 10^4$. From 'Propagation of Short Radio Waves' by D. E. Kerr, copyright 1951, McGraw-Hill Book Co. Used by permission

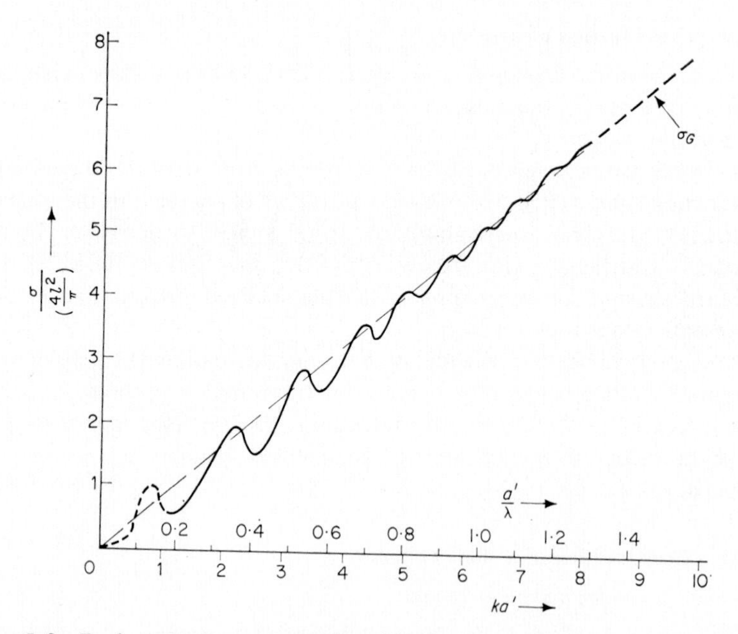

Fig. 5.9 Back-scattering cross-section (σ) of a short thick perfectly reflecting cylinder in broadside aspect in the electromagnetic case (after Mentzer)[9]

in the electromagnetic case of transverse electric field, which corresponds to a rigid cylinder in acoustics. The energy back-scattered by the cylinder in broadside aspect in the geometrical region is proportional to its radius a', to the frequency and to the square of its length l.

The effect of change of radius of cylinder a' is seen in Fig. 5.9 in which the ordinate is a summation representing the back-scattering cross-section per unit length of cylinder in broadside aspect, multiplied by $\pi/4$. This graph shows regular fluctuations which become more pronounced when the Rayleigh region is approached. In the geometrical region, the graph is asymptotic to the value $\frac{1}{4}\pi ka'$ (shown by the broken straight line) when the cross section is

$$\sigma_G = ka'l^2 \tag{5.3}$$

(see Table 5.1). In the case of a partly-reflecting material, this equation becomes:

$$\sigma_G = ka'l^2\mathscr{R}^2 \tag{5.4}$$

considering the front interface only. In the Rayleigh scattering region (below $ka' = 1$), σ falls according to a fourth-power law of a'/λ (when the length of the cylinder is also in this region).

When the axis of the cylinder is inclined (Table 5.1), the general formula indicates a back-scattering polar diagram dependent on the angular deviation from broadside aspect (ϕ). A part of the calculated polar diagram near broadside aspect for a rigid cylinder of length 16·7 λ and radius 1·58 λ, is seen in Fig. 5.10 and is compared with the experimental measurements using a cylinder of these dimensions made of steel. Agreement is close, indicating that steel acts almost like a rigid material.

As has been seen, in the geometrical region the acoustic cross-section of a cylinder is proportional to the square of its length (l) since $\sigma_G = ka'l^2$ but when its radius approaches the Rayleigh scattering region, a regular fluctuation occurs (Fig. 5.9). A similar effect also occurs when the length approaches this region. Fig. 5.11 (full line) gives the author's experimental results[14] for 44 copper cylinders of various lengths but all of radius $a' = 0{\cdot}20\ \lambda$, taken at 360 kHz. (Copper is almost perfectly reflecting to sound, $\mathscr{R} = 93$ per cent.) The echo voltage in broadside aspect is compared with that calculated to come from a target of cross-section σ_G, shown by the horizontal broken line. At $l = 4{\cdot}9\ \lambda$, the difference between the measured and calculated values is within the experimental error.

At small values of l/λ, this graph indicates a value of σ proportional to the square of l/λ since a' is still in the geometrical region.

Some results due to Sothcott[15] using cylinders made from expanded polyvinyl chloride (that is, substantially air-filled) taken at 15 and 30 kHz,

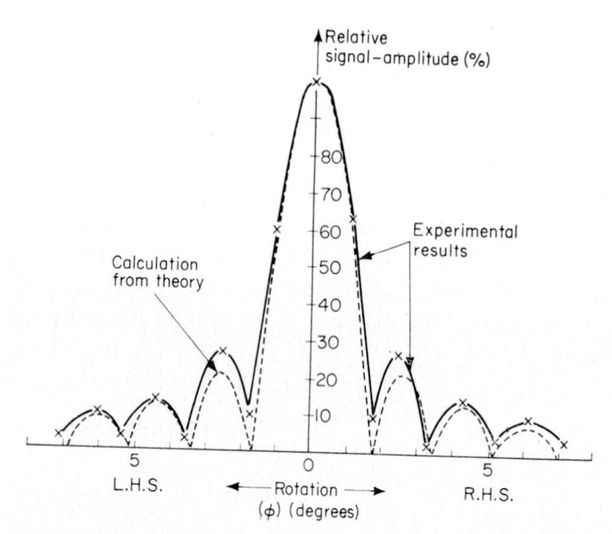

Fig. 5.10 The signals received from a short steel cylinder of length 16·7 λ and radius 1·58 λ at 1·5 MHz (λ=1·00 mm), compared with those calculated for a rigid cylinder having the same dimensions[8]. (For broadside aspect $\phi=0$)

have also been added to Fig. 5.11 for comparison and are seen to confirm the fluctuations for values of l/λ between 1 and 5.

5.11.2 Cylinder translucent to sound

The mathematical treatment was extended further by Faran[16] to infinite cylinders of solid materials which transmit sound, having various radii, taking into account the additional effects of shear waves. His experimental results confirmed the existence of sharp minima in the back-scattered sound at radii near those at which normal modes of free vibration of the cylinder occur, which satisfy the conditions of symmetry, that is when

$$k_1 a' = 1{\cdot}18,\ 1{\cdot}43,\ 1{\cdot}81,\ 2{\cdot}17,\ 2{\cdot}25,\ 2{\cdot}36,\ 3{\cdot}01, \text{ etc.} \qquad (5.5)$$

where $k_1 = 2\pi/\lambda_1$ and λ_1 is the wavelength of compressional waves in the material of the cylinder.

Part of a polar diagram of a short thick cylinder of Perspex having its length in the geometrical region is seen in Fig. 5.12, in which the measured signals (at 360 kHz) are found to exceed those calculated for the front water–Perspex interface only, using Eqn. 5.4 of Section 5.11.1. Perspex is a material which is only partly reflecting ($\mathscr{R}$=36·5 per cent) so that the rear Perspex–water interface contributes to the overall echo. At this particular

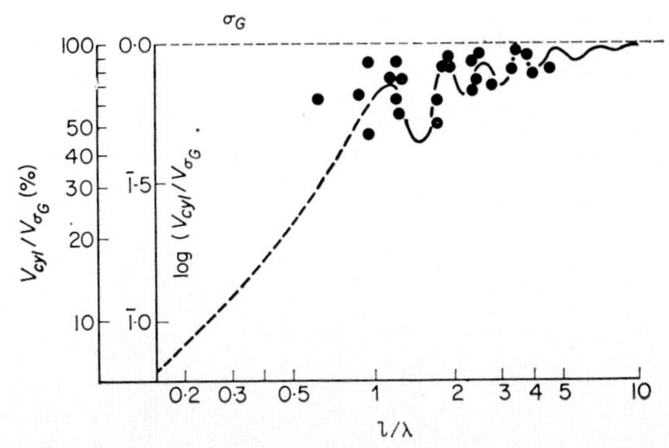

Fig. 5.11 Variation of the voltage ratio between the measured signal from the entire cylinder (V_{cyl}) and calculated signal from the front surface only in the geometrical region ($V\sigma_G$), for different lengths of cylinder[14]. ● Sothcott's readings for expanded polyvinyl chloride cylinders, for comparison

radius of cylinder ($a'/\lambda = 0{\cdot}33$), the amplitude is increased by a factor of 3 in the main lobe and less in the side lobes, but the angular positions of the first few minima on each side are not greatly affected.

The effect of cylinder radius on the acoustic cross-section per unit length for materials of various degrees of translucence to sound, appears in Fig. 5.13. (The length of the cylinder is assumed to be in the geometrical region.) These measurements were also taken at 360 kHz ($\lambda = 0{\cdot}41$ cm in water), and are compared with calculated values of the cross section of the front surface only, in the geometrical region.

Regarding the effect of cylinder radius in the Rayleigh scattering region,

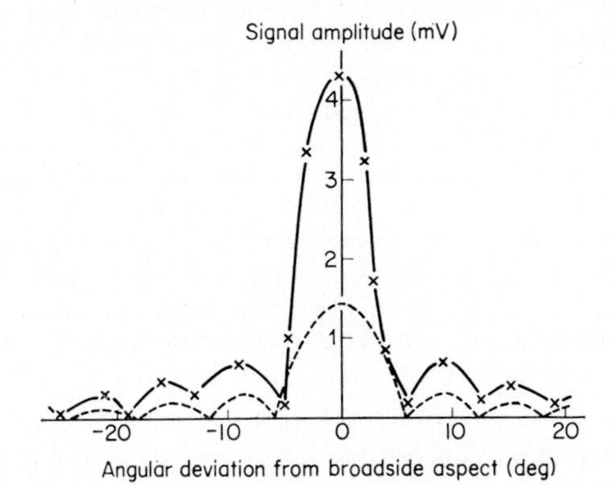

Fig. 5.12 Signals back-scattered by a short Perspex cylinder of length 2·0 cm ($4{\cdot}9\lambda$) and radius 0·1345 cm ($0{\cdot}33\lambda$) (full line) compared with those calculated for the front surface only, of reflectivity $\mathscr{R} = 36{\cdot}5$ per cent (broken line)[14]

the cross section of the entire cylinder falls below σ_G, increasingly so as a' diminishes. A further decline occurs if a material of lower reflectivity is used.

When the radius of the cylinder is increased so that it begins to enter the geometrical region, undulations in signal occur, which may be regarded as

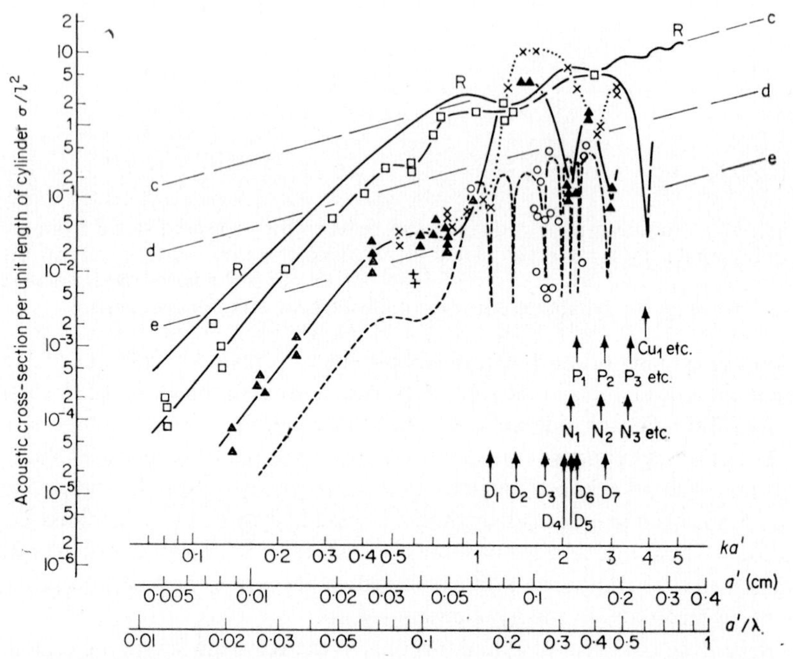

Fig. 5.13 Signals back-scattered by various cylinders in broadside aspect plotted against radius of cylinder a'. All the cylinders are of length 2 cm[14].

Experimental points; □ (full line), Copper (Cu), $\mathscr{R}=93$ per cent; × (dotted line), Perspex (P), $\mathscr{R}=36{\cdot}5$ per cent; ▲ (full line), nylon (N) $\mathscr{R}=35{\cdot}3$ per cent; ○ (broken line), Dawn rubber (D), $\mathscr{R}=11$ per cent; + fish bone (F), $\mathscr{R}=26$ per cent. The graph for the rigid cylinder (R, full line) $\mathscr{R}=100$ per cent is transferred from Fig. 5.9.

The straight lines indicate the acoustic cross-section of the front surface only in each case calculated from Eq. 4; cc, rigid, $\mathscr{R}=100$ per cent; dd, nylon, $\mathscr{R}=35{\cdot}3$ per cent; ee, Dawn rubber, $\mathscr{R}=11$ per cent. The values of ka' at which the first few minima occur (calculated after Faran), are shown by arrows

the effect of interference between the signals scattered by front and rear surfaces of the cylinder. This first occurs at $ka'\simeq 1$ and continues into the geometrical region. At this maximum, the cross section for the entire cylinder can exceed σ_G by a considerable amount. The excess is small in the case of the rigid cylinder ($\mathscr{R}=100$ per cent) but can be as much as a 23/1 increase over σ_G for Perspex where a large proportion of the incident sound penetrates the material. For low-reflectivity materials, for example 'Dawn' rubber, σ can be as large as 8 times σ_G.

Sharp minima were found in the back-scattered energy, as predicted by Faran. The positions of these minima depend on the velocity of sound in the material and have an over-riding effect on the relationship between σ and ka', giving a very complex graph.

By interpolation in Fig. 5.13, an approximate graph may be obtained for any material having a density greater than that of water, when the velocity of compressional sound waves in the material is known.

5.11.3 *Cylinders with coaxial cylindrical holes*

The situation becomes more complicated when the cylinder has an air-filled coaxial cylindrical hole extending along its whole length, with the ends sealed. This structure applies to a number of targets but is particularly applicable to a fish which has a gas-filled swim-bladder within the body of flesh (see Section 5.20).

Experimental measurements at 1·5 MHz have been made in some detail on short rubber cylinders having cylindrical holes of approximately one third the diameter of the cylinder[17]. Before sealing the ends of the holes with thin adhesive tape, the dimensions of the rubber cylinders and holes were measured under a travelling microscope at each end in two directions at right angles and the mean values obtained. Using the scale-model apparatus, the pulse had a duration of 25 μs or 37 λ in water. For the sizes of cylinders and for the particular rubber of low reflectivity which were used ($\mathscr{R}=11$ per cent), it may be shown that the contributions inter-reflected between the surfaces of the cylinder and of the hole, die away quickly so that this length of pulse give a result which differs but slightly from that for a continuous wave.

Comparisons were first made between the back-scattering polar diagrams of a number of cylinders without hole against those of cylinders of the same dimensions but with the hole. Typical results are seen in Figs. 5.14 and 5.15. (Signals from the ends of the cylinders were ignored.) On examination of Fig. 5.14, it is clear that at a certain value of cylinder diameter D and of thickness of rubber t (between front surface of cylinder and front surface of hole), little change in signal level occurs when the hole is introduced. On the other hand, Fig. 5.15 shows that at other values of D and t, there is a marked change in signal level on the introduction of the hole. (The lobes in Fig. 5.14 for a longer cylinder are also packed more closely together than in Fig. 5.15.) By comparison with a standard target, the measured acoustic cross-sections may be calculated.

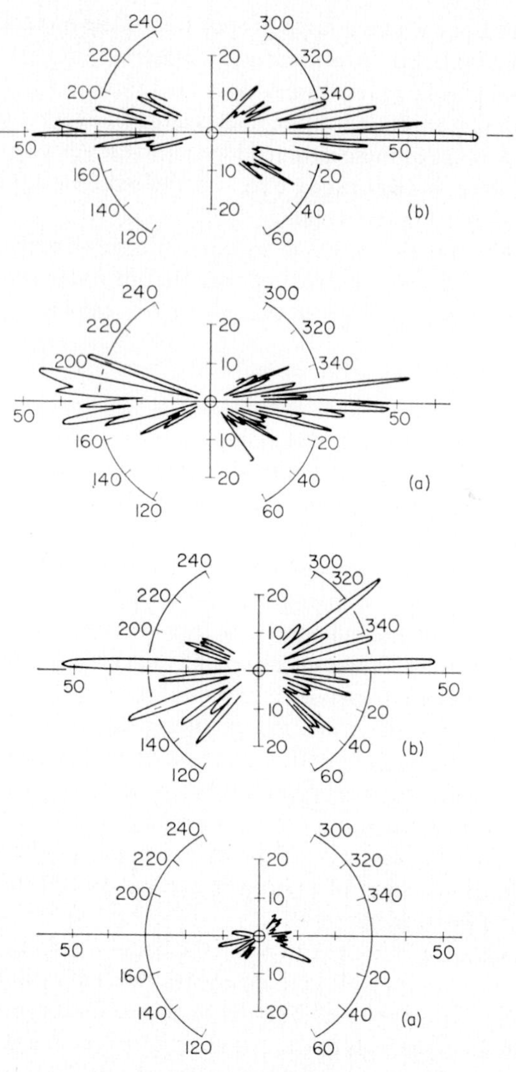

Figs. 5.14 and 5.15 Typical back-scattering polar diagrams of rubber cylinders with and without coaxial air-filled cylindrical holes. The right-hand abscissa corresponds to broadside aspect and angular rotation from this position ϕ is given in degrees. Radial coordinate: mV. (a) Without hole, (b) with hole. Right-hand side: front; left-hand side: rear.
Dimensions:

Fig. 5.14: Length $l=1{\cdot}327$ cm, diameter of cylinder $D=0{\cdot}440$ cm, diameter of hole $d'=0{\cdot}138$ cm and thickness of rubber $t=1{\cdot}51$ mm.

Fig. 5.15: $l=0{\cdot}442$ cm, $D=0{\cdot}437$ cm, $d'=0{\cdot}123$ cm and $t=1{\cdot}57$ mm.

The readings in broadside aspect may be analysed further. The effect of length of cylinder is eliminated by dividing by the square of length and the value of σ/l^2 so obtained may be compared with that calculated for the front surface of the cylindrical hole alone (which, being filled with air, has a cross-section larger than that of the rubber cylinder alone). Fig. 5.16 gives the

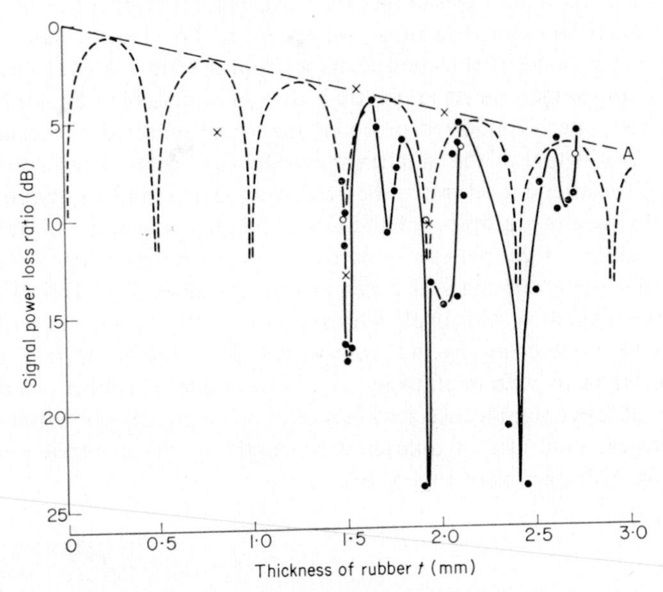

Fig. 5.16 Cylinders with holes: signal power loss ratio in broadside aspect plotted against thickness of rubber. Ordinary cylinders ×; cylinder with hole, placed eccentrically: ○ first method with polar diagrams, ● second method giving maxima and minima when adjusting t continuously. (The line indicates the mean of the two sets of readings.) Alongside each point, the corresponding example number is given, where applicable.

Also shown (broken line) is a tentative graph corresponding to a parallel-sided plate made of rubber[17]

ratio between these measured and calculated values, plotted against thickness of rubber t. A second series of readings were taken (using a single cylinder with the hole placed eccentrically), concentrated on the maximum and minimum values of the broadside signal when the cylinder was rotated slowly about its longitudinal axis. In this way, a continuous thickness of rubber t between 1·49 and 2·73 mm was examined and no sharp minima could be missed. Care was exercised to ensure that the cylinder was precisely in broadside aspect throughout.

As seen in Fig. 5.16, all the signals fall below that predicted for the hole alone and the peak values indicate a decline along the line OA which may be attributed to the acoustic attenuation of the rubber.

On a simple theory using geometrical acoustics, two 'highlights' might be expected, one of the convex surface of the rubber, the other on the convex surface of the hole, and these would interfere to give minima at values of t which are integral multiples of half the wavelength in rubber ($\lambda = 0{\cdot}92$ mm).

The effects of inter-reflections are complex. To give an indication, a parallel-sided plate of thickness t made from a solid material having an acoustic impedance much larger than that of water, may be considered. When this plate is immersed in water, the signal reflected in normal incidence varies with t in the manner shown by the broken line in Fig. 5.16 allowing for attenuation in the material, with sharp minima at values of t which are integral multiples of half the wavelength in the material. When the water backing to the plate is replaced by air, in practice, the author has found that these minima still occur at similar values of t. This effect has been investigated by Aldridge[18] who suggested a theory which ascribes the minima to mode-changing to Lamb waves. Thus the graph in Fig. 5.16 may be taken to refer tentatively to a plate made of rubber. Additional minima of lesser significance at values of t approximately corresponding to odd integral multiples of quarter wavelengths in the material were also found by Aldridge (as in Fig. 5.16).

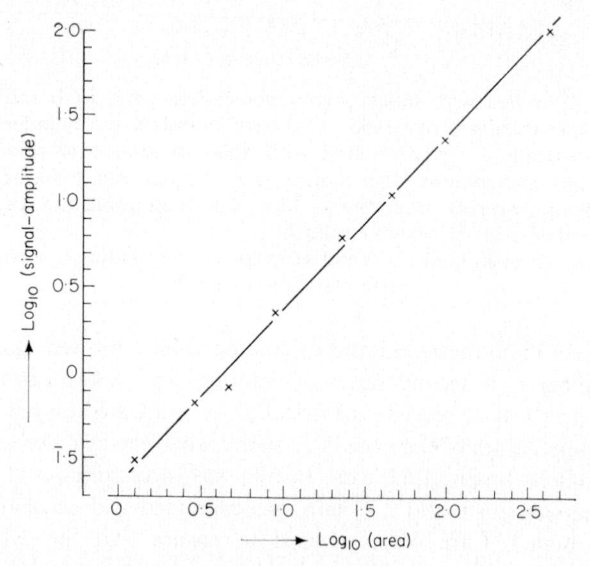

Fig. 5.17 Signal amplitudes received from plane targets of various areas (at constant range)[8]

Fig. 5.14 is now seen to approximate to a deep minimum when the presence of the hole does not appear to significantly affect the overall signal.

It may be concluded that in the case of a cylinder made from a material of low acoustic reflectivity in which there is an air-filled hole, simple calculations based on geometrical acoustics assuming two 'highlights' prove satisfactory as a first approximation when the contribution from the hole predominates.

The effect of the introduction of the hole is to make the back-scattering polar diagram more omni-directional than that of a single rigid interface alone (as given by Table 5.1), although the maximum in broadside aspect remains unaltered at the value calculated for the front surface of the hole, after allowing for transmission through the water/body interface there and back and for attenuation in the body material from interface to hole and back. The number of lobes in the complete back-scattering polar diagram is always less than for a perfectly-reflecting cylindrical hole and falls at the minima in Fig. 5.16.

5.12 Cross sections of plane targets

5.12.1 Plane targets opaque to sound

As seen in Table 5.1, in the geometrical region the cross section of a plane interface in broadside aspect increases according to the square of frequency and the square of its area. The latter may be confirmed, for example, in a scale-model experiment in which a number of fresh dry pieces of thin cork sheet were cut and, in turn, orientated to lie perpendicular to and on the axis of the beam at constant range in the far-field region. (Their dimensions were measured under a travelling microscope.) Cork approximates to a perfectly-reflecting surface. The amplitudes of the resulting echoes were observed (at 1·5 MHz) and are seen in Fig. 5.17 in which the slope of the best straight line through the points is 1·00, indicating a cross section increasing as the square of area.

5.12.2 Plane targets translucent to sound

Rayleigh gave the form of the total amplitude reflection coefficient $\mathscr{R}_T$ in the case of a parallel-sided plate of thickness d and acoustic impedance Z_2, separating two regions of impedance Z_1

$$\mathscr{R}_T = \left(\frac{Z_2}{Z_1} - \frac{Z_1}{Z_2}\right) \Big/ \left[4 \cot^2 \frac{2\pi d}{\lambda p} + \left(\frac{Z_2}{Z_1} + \frac{Z_1}{Z_2}\right)^2\right]^{1/2} \tag{5.6}$$

where λp is the wavelength of sound in the plate.

This formula refers to continuous longitudinal waves in normal incidence and takes account of both the contributions from the front and rear interfaces. The values of $\mathscr{R}_T$ have been calculated for different values of the ratio Z_2/Z_1 and are plotted against d in Fig. 5.18[11]. In the absence of attenuation in the plate, the graphs would repeat within each interval of 0·5 in the value of $d/\lambda p$. These results would be valid for pulses, provided the pulse is much longer than twice the time for the sound to cross the plate. The full lines show the graphs for various materials having values of $Z_2/Z_1 = 1{\cdot}1$, 1·2, 1·5, 2·0, 4·0 and 32.

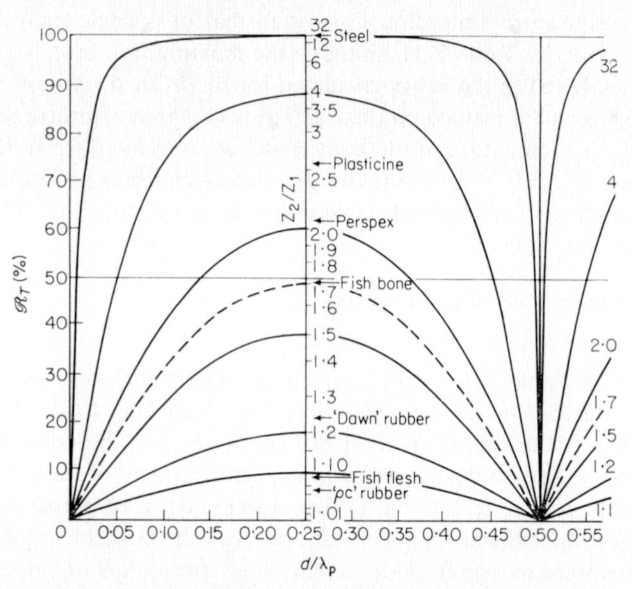

Fig. 5.18 Reflection of continuous waves from a plate of acoustic impedance Z_2 and thickness d, immersed in a medium of impedance Z_1, in normal incidence (neglecting attenuation in the plate)[11]

The number of inter-reflections which contribute significantly to the total echo may be studied by taking the example of a plate of Perspex for which the reflectivity (at one interface) $\mathscr{R} = 35{\cdot}3$ per cent. It is possible to calculate the relative energies of the various components reflected by a parallel-sided plate of this material immersed in water (Fig. 5.19)[11], neglecting attenuation in the first instance. The overall strength of the echo is obtained by adding the *amplitudes* of each of these energy components in correct phase, bearing in mind that the rear surface introduces a phase-change of 180° since it acts like a pressure-release surface ($Z_2 > Z_1$). For example, when $d = 5\lambda/4$ the phases are as follows

$$A+;\ B+;\ C-;\ D+;\ \text{etc.}$$

A calculation based on these four components, only, gives a value of $\mathscr{R}_T = 62{\cdot}8$ per cent without allowing for attenuation (which agrees closely with that from Eqn. 5.6).

Attenuation in the Perspex may then be allowed for by reducing each of these components by the correct amount related to the total path in the plate (taking the coefficient of attenuation in Perspex as 3·4 dB cm^{-1}). After allowing for attenuation, $\mathscr{R}_T = 58{\cdot}6$ per cent when $d = 5\lambda/4$. The calculated effect of attenuation in the plates of various thicknesses is seen in Fig. 5.20

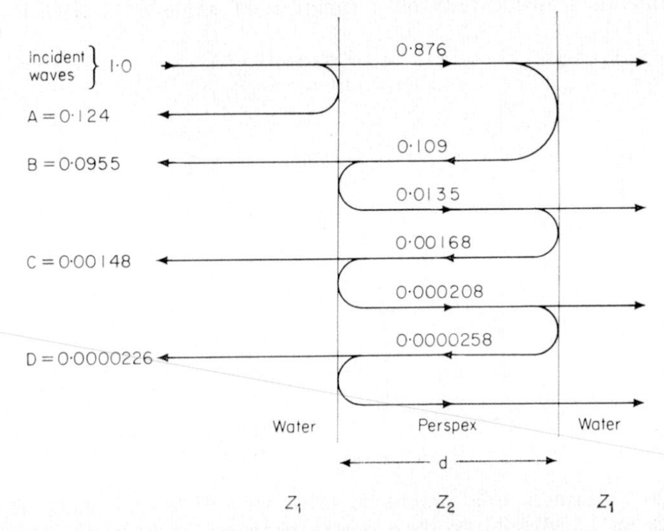

Fig. 5.19 Relative energies of the various components reflected by a parallel-sided Perspex plate immersed in water (normal incidence). ($Z_2/Z_1 = 2{\cdot}09$; $\mathscr{R} = 35{\cdot}3$ per cent)[11]

(broken line). The peak values of $\mathscr{R}_T$ decline as the plate thickness increases. At the same time, the minimum values increase. (For an infinitely thick plate having zero contribution from the rear interface, the loci of peaks and minima, shown by the horizontal lines would, of course, converge to a value of $\mathscr{R}_T = \mathscr{R} = 35{\cdot}3$ per cent.)

These calculations are confirmed by experimental measurements made at 1·5 MHz on a number of parallel-sided Perspex plates each of dimensions 2·91 mm × 7·07 mm but of various thicknesses (also seen in Fig. 5.20). (The absolute value of $\mathscr{R}_T$ was determined by comparison with a standard target.) In view of the agreement between the calculated and measured values, it is clear that scattering through the side walls of the plate (i.e., at right angles to the incident direction) is not significant, also that the amount of mode changing (e.g., to shear waves) is small.

So far in this Section, only normal incidence has been considered. Turning now to other angles of incidence (ϕ), Fay and Fortier[20] have measured the transmittivity of a steel plate immersed in water using pulses 20 μs long, i.e., long enough to allow a steady state to be built up in the plate. A projector was placed on one side of the plate and a hydrophone on the other. Readings were made with the test plate at various orientations and compared with the straight-through signal with the plate removed, to determine its transmittivity as a function of angle of incidence ϕ. The

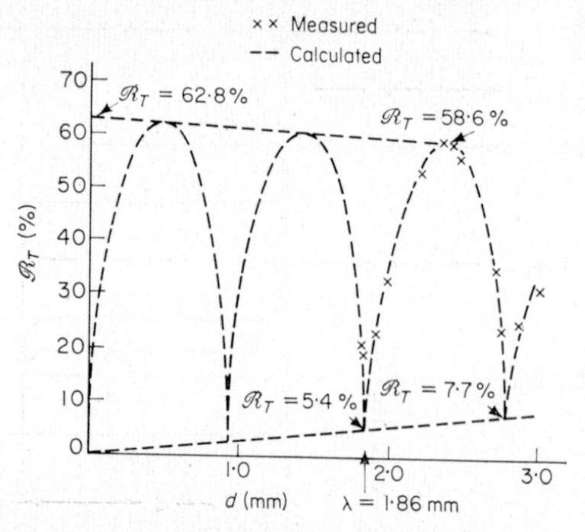

Fig. 5.20 Measured total amplitude reflectivity of Perspex plates in normal incidence (at 1·5 MHz)[11]. As the acoustic characteristics of Perspex are known, these results may be compared with the calculated values assuming $Z_2/Z_1 = 2{\cdot}09$, $\lambda = 1{\cdot}86$ mm and coefficient of attenuation = 3·4 dB cm^{-1}. Readings taken at constant range

transmittivity depends also on the product of frequency and plate thickness. The frequency was constant at 1·5 MHz. Measurements were made at a large number of thicknesses of plate, by grinding a few thousandths of an inch from it between each set of readings.

The results are seen in the form of a three-dimensional graph in Fig. 5.21 which consisted of an assembly of cut-out graphs taken at each thickness. This graph must be viewed with reference to a rectangular box consisting of three coordinates. The horizontal coordinates are the angle of incidence (ϕ) across the graph on each side of the centre-line and the product fd (frequency × plate thickness) from front to rear. The vertical coordinate is the transmission loss which increases in the downward direction.

A three-quarter view appears in (a) and a front elevated view in (b). The positions of the 'mountain ranges' represent values of ϕ and fd at which the transmission losses are small. The broad ridges, which appear only at angles of incidence exceeding about 35°, are attributed to diffraction. The minimum in transmittivity in normal incidence at $d=\lambda/4$ (compare with Fig. 5.18 which shows *reflectivity*) is point A and the maximum at $d=\lambda/2$ is point B.

A plan view of these 'mountain ranges' is shown in Fig. 5.21(c). The left-hand half of this diagram may be compared with that given by Fay and Fortier (d) in which the scale in fd has been compressed. The latter shows the calculated positions (full line) and their experimental results (dots).

Fig. 5.21(e) gives a vertical slice through the three-dimensional graph in which all values of ϕ have been condensed. This figure indicates the observed losses at various points on the 'mountain ranges' in terms of fd. (Each 'mountain range' in (c), (d) and (e) has been given a Roman numeral to aid cross-reference.)

5.13 The Underwater Acoustic Camera

A high-resolution continuous-wave system has been developed to give acoustic 'pictures' of underwater targets[3]. The method may be likened to that of closed-circuit television employing lights, camera and monitoring equipment. However, here the comparison ends, for, in practice, the apparatus turns out to be widely different because the wavelength is relatively large (about 2000 times greater), also the acoustic image must be converted to an electronic signal at high speed (as there is no storage of signal in the camera tube) and since the equipment is used under water. The benefit of using ultrasonic waves rather than light, is the greater penetration in turbid water.

The equipment was mounted on a moored barge (Fig. 5.22). The target was supported on ropes and was continuously irradiated with ultrasound at 1·2 MHz by the transmitting transducer. A real acoustic image of the target was formed by a concave spherical mirror (of diameter 81 cm) and was transformed to a television picture by means of a special acoustic to electronic image-converter tube[21]. The combination of the mirror and converter tube acted rather like an acoustic television 'camera' having a field of view of about $\pm 2\frac{1}{2}°$. The transmitting transducer and the 'camera' assembly were mounted on training shafts allowing pan and tilt.

The resulting pictures were viewed on the monitor tube below deck. When examining these pictures, it must be borne in mind that the acuity of the system is about 0·1° (being related to the Airy disc in the image plane

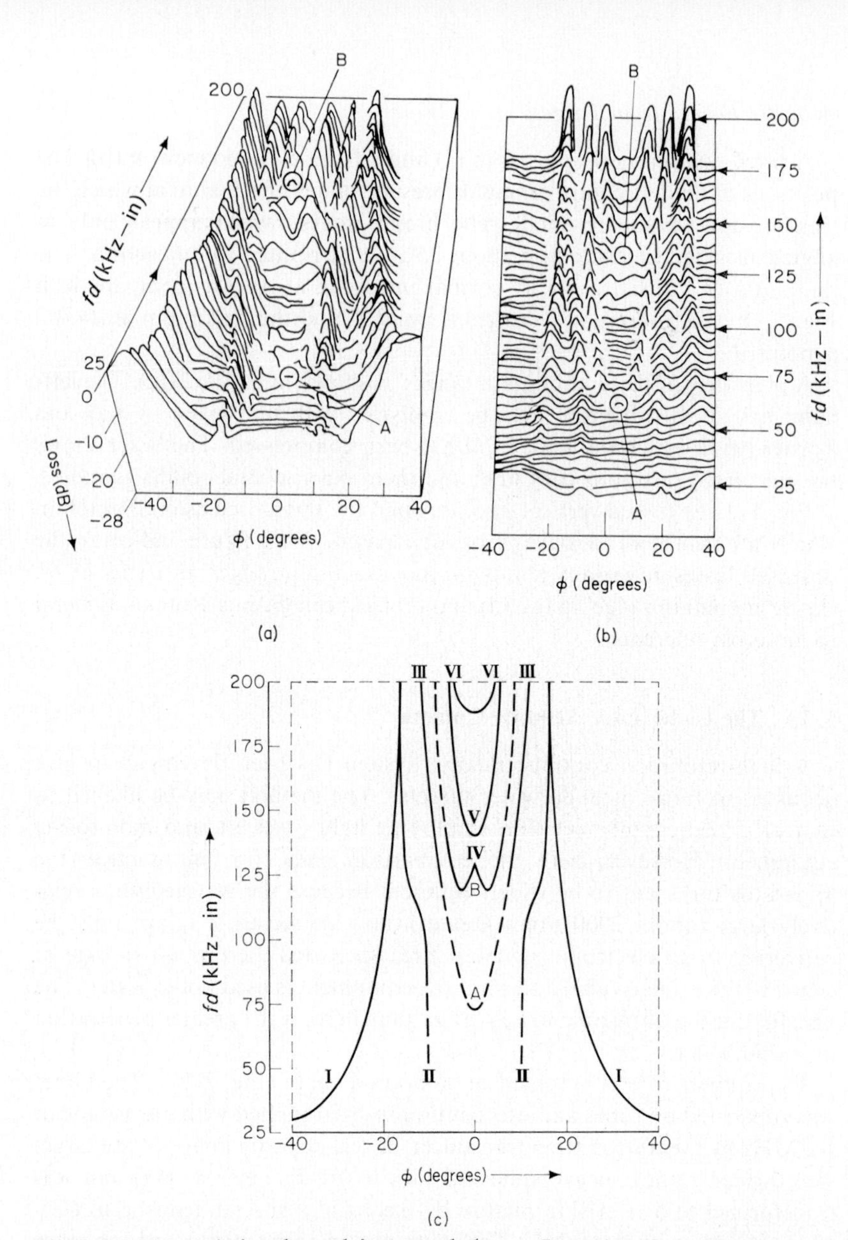

Fig. 5.21 Transmission through immersed plates—Intensity of transmitted sound observed as a function of angle of incidence (ϕ) and product of frequency (f) and plate thickness (d)[20].

A minimum at $d = \lambda_p/4 \equiv fd = 59$
B maximum at $d = \lambda_p/2 \equiv fd = 118$

(a), (b) Two views of the three-dimensional graph.
(c) Plan view showing positions of 'peaks'.

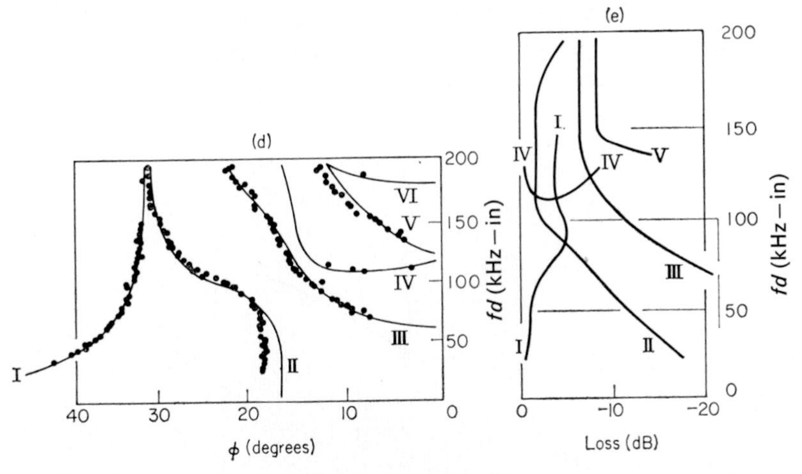

Fig. 5.21 (*continued*)

(d) Plan view of left-hand half of model. Continuous curves calculated from classical equation. Observed 'peaks' plotted as dots.

(e) Observed values of losses at the 'peaks' in the model, plotted as smooth curves.

which is formed by a 'point' object at infinite range). The coherent nature of the ultrasound, having a wavelength of 0·12 cm, also causes interference effects. In an alternative mode of operation, these can be randomized by varying the transmitted frequency by ± 10 kHz, about 33 cycles of frequency modulation occurring per second.

5.13.1 Appearance of target on the Underwater Acoustic Camera

Some examples of the appearances of a number of targets when irradiated with ultrasonic waves at 1·2 MHz are seen in Fig. 5.23. Some of these targets were specular reflectors and others were rough. In all cases when using the fixed frequency, the pictures scintillated owing to the movement of the target against the general cross-talk background. This movement was difficult to prevent and gave a noticeable effect on the picture, since the wavelength is so short (0·12 cm). On switching to frequency modulation, these interference effects were eliminated and the main highlight was seen, giving a better impression of shape of target.

A sphere (a) gave a very characteristic picture, consisting of a series of concentric rings (b) with intensity falling rapidly from the centre. The scintillation took the form of radial movement of these rings, inwards or outwards. On FM a single highlight (c) was seen.

In the case of a hollow cylinder (d), the rings became ellipses on CW (e) from which it was possible to tell the attitude of the cylinder. The main distribution of intensity was along the major axis of the ellipse, which lay in

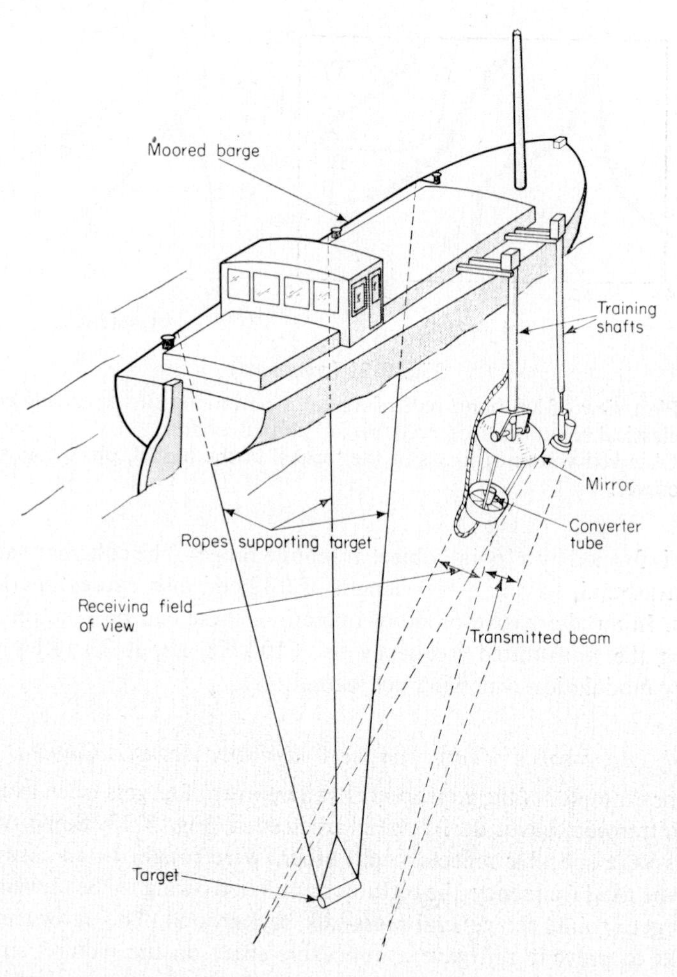

Fig. 5.22 Arrangement of the Underwater Acoustic Camera on a moored barge. The electronic control room and picture monitor were below deck[3]

the direction of the length of the cylinder. The more conventional picture of the highlight (f) appeared on FM. It was not possible to irradiate the whole length of the cylinder uniformly, owing to the restricted beam-width of the transducer. This applies to many of the targets and in the corresponding pictures, only a portion of each target is 'illuminated'.

Another factor which assists the observer in recognizing the shape of the target is motion relative to the Underwater Acoustic Camera.

Other specular reflectors were tried. For example, the corner of a square plate (g) gave a complex interference pattern (h). All the plane specular targets needed to be very carefully adjusted in aspect.

Turning to rough targets, (k) is the picture obtained when the rock (j) was suspended in the field of view, whilst (l) shows the sea bed, which consisted of mud, shingle and debris. The picture of the sea bed had a characteristic fixed irregular granular structure. That of the rock had the same structure but was sufficiently pronounced almost to suggest shape. Both scintillated with CW but with FM the general shape of the rock was seen to move slightly as the rock swung on the ropes.

5.14 Echoes from fish

The back-scattering of acoustic waves by such underwater targets as fish may be discussed more freely than the case of security classified targets. However, an analysis of the acoustic behaviour of fish indicates the principles which apply to all types of targets.

Fig. 5.24[22, 23] shows the echo from a single fish as seen on an expanded CRT display on a vertical echo-sounder (as in Fig. 5.1). This display readily allows the amplitudes and shapes of the fish echoes to be observed and the expanded time-base, which may be locked to the sea-bed echo, is particularly suitable for sea-bed trawling. The transmitted pulse at 30 kHz was about 0·5 ms long (between half-amplitude points) and the shape of the echo is determined by the overall bandwidth of the system (approximately 2 kHz). The degree of range acuity is about 15 in so that the echoes from the separate parts within a fish cannot be distinguished. Some random noise is superimposed on the time base.

5.15 Echoes from the sea bed

At the foot of Fig. 5.24 are seen the first few half-cycles of the initial rise of the sea-bed echo. Since the latter is a much larger target than the fish, the sea-bed echo has an amplitude about 300 to 1000 times larger than the fish echo. Thus, in vertical sounding, fish echoes at slant ranges greater than the shortest path to the sea bed, disappear into the sea-bed echo.

A typical sea-bed echo is seen in Fig. 5.25 on a longer time-base and at greatly reduced receiving amplifier gain as compared with Fig. 5.24. The length of the sea-bed echo depends on its acoustic roughness and on the

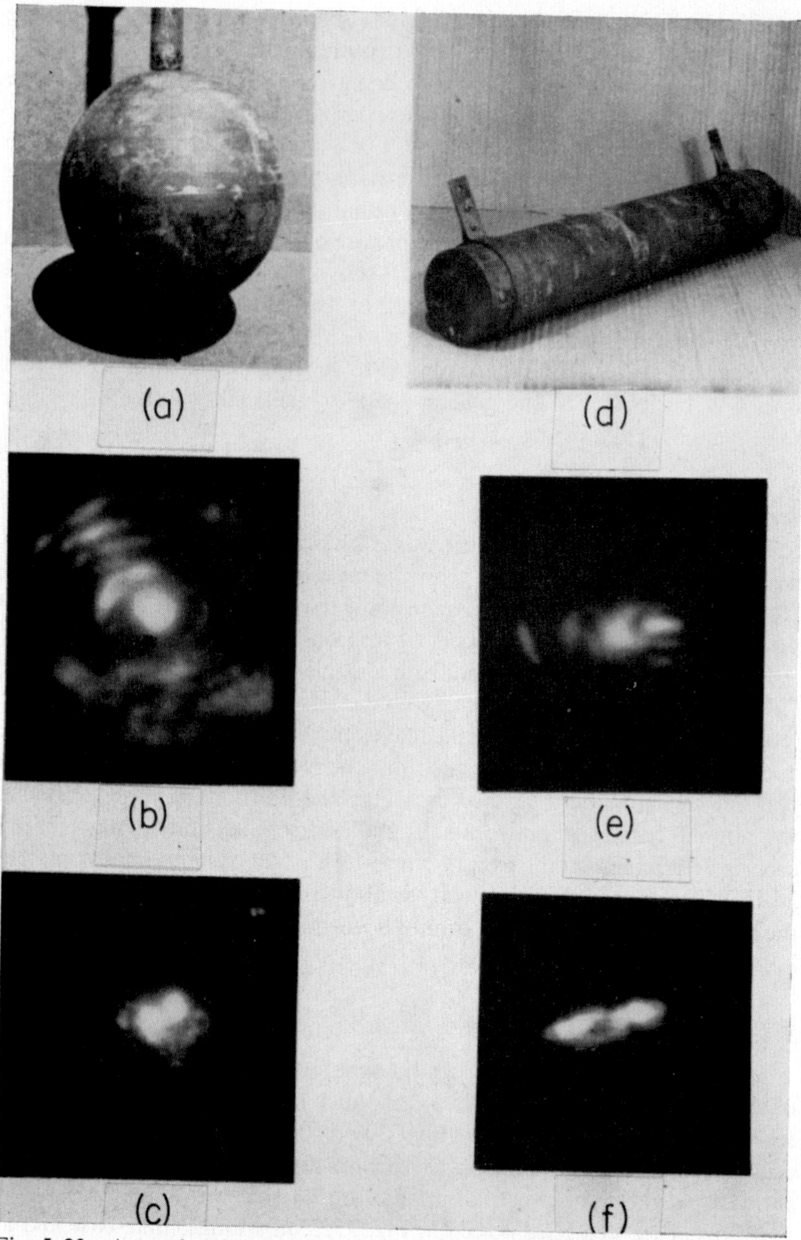

Fig. 5.23 Acoustic picture of various targets irradiated with ultrasound at 1·2 MHz[3], compared with normal views.

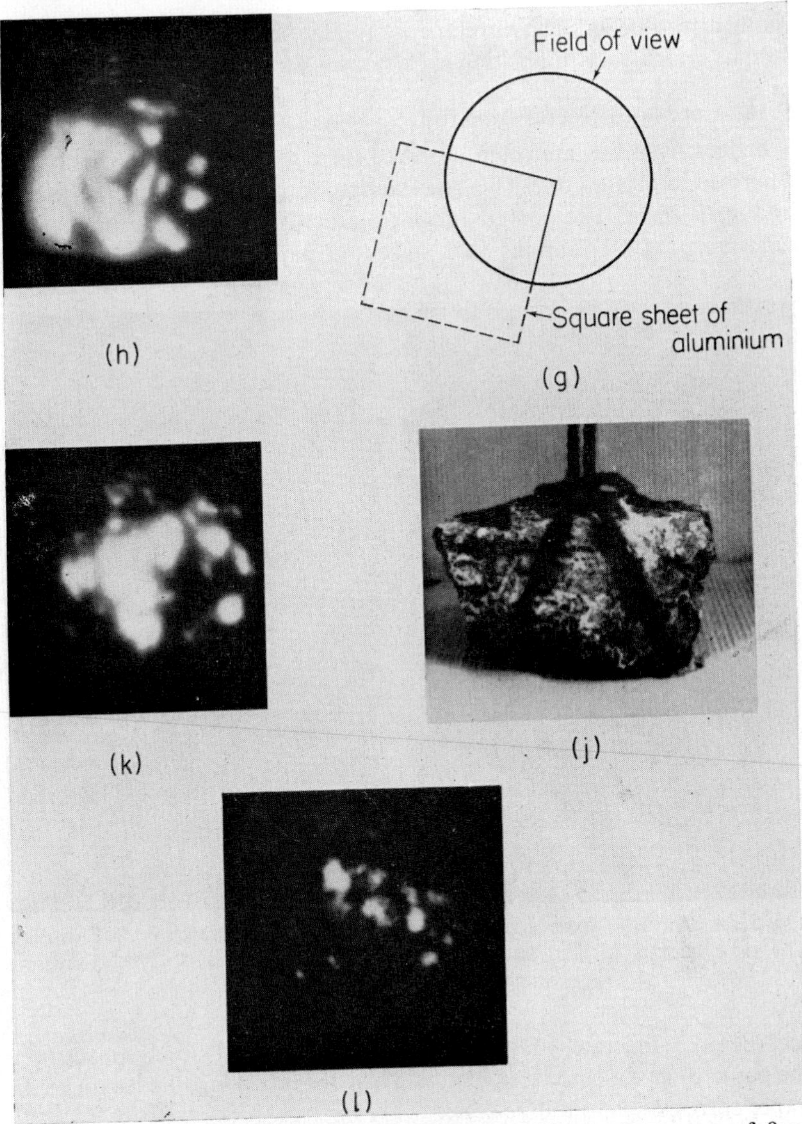

(a), (b), (c) Air-filled metal sphere of diameter 35 cm at a range of 9 m; (b) using CW; (c) using FM.
(d), (e), (f) Air-filled metal cylinder, 10 cm diameter and 60 cm long at 9 m; (e) using CW; (f) using FM.
(g), (h) Corner of a brass plate 45 cm square and 3 mm thick at 5·4 m; CW.
(j), (k) A rock 30 cm wide, suspended on a rope, at 5·4 m, CW.
(l) Sea bed under the barge at a range of 9 m, mud, shingle and debris, CW.

width of the beam. The envelope is usually broken up by a number of minima which vary in position each transmission 'ping'.

5.16 Correlated records from fish

Echoes from fish are often observed on a recorder of the type already described in Section 5.2. This gives an intensity-modulated record as seen in Fig. 5.26(b). The curved range-scale across the paper represents an echo-range of 60 fathoms and since the paper is driven through the

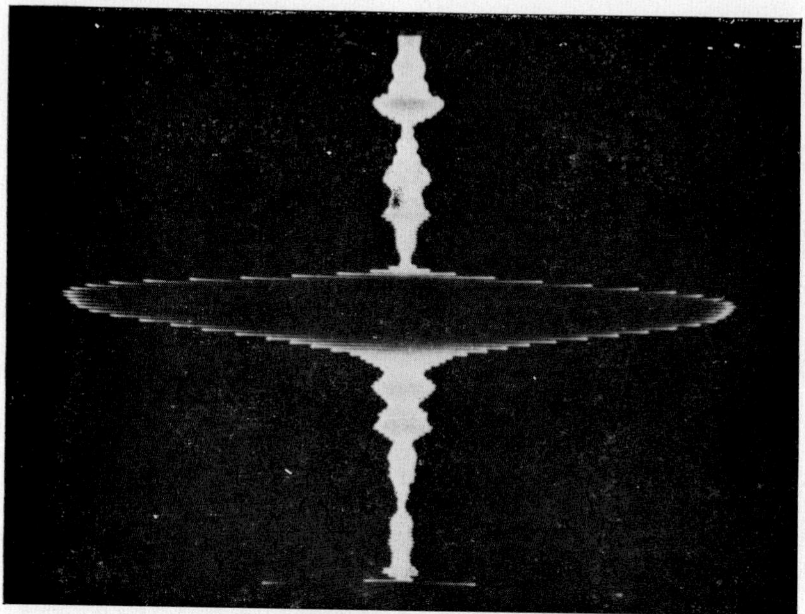

Fig. 5.24 An echo from a single fish as seen on an expanded CRT display. The time base is vertical and represents a range annulus of 2 echo-fathoms (range increasing downwards)[23]

recorder at a slow rate, consecutive 'pings' are recorded closely adjacent on the paper to give a correlated record. Thus distance along the length of the paper corresponds to a time scale.

Echoes from fish are seen just above the sea-bed echo. As the ship moves along, a fish is detected initially at the edge of the acoustic beam (which is fairly wide), subsequently at the centre of the beam at a shorter range and finally at the opposite edge of the beam at the original range. During this process which occupies several 'pings', a characteristic fish trace is drawn on the record in the form of a half-hyperbola (or 'comet').

Fig. 5.26(b) also shows the use of the 'white-line' technique whereby after the initial rise of the sea-bed echo is received, the receiving amplifier is blanked for a short period. This assists in the detection of fish echoes very near the sea-bed echo.

Observation of fish near the sea bed is also greatly assisted by the use of a recording scale-expander[24] (which is a mechanical analogue of the expanded CRT display) but gives an intensity-modulated record. In this case, the

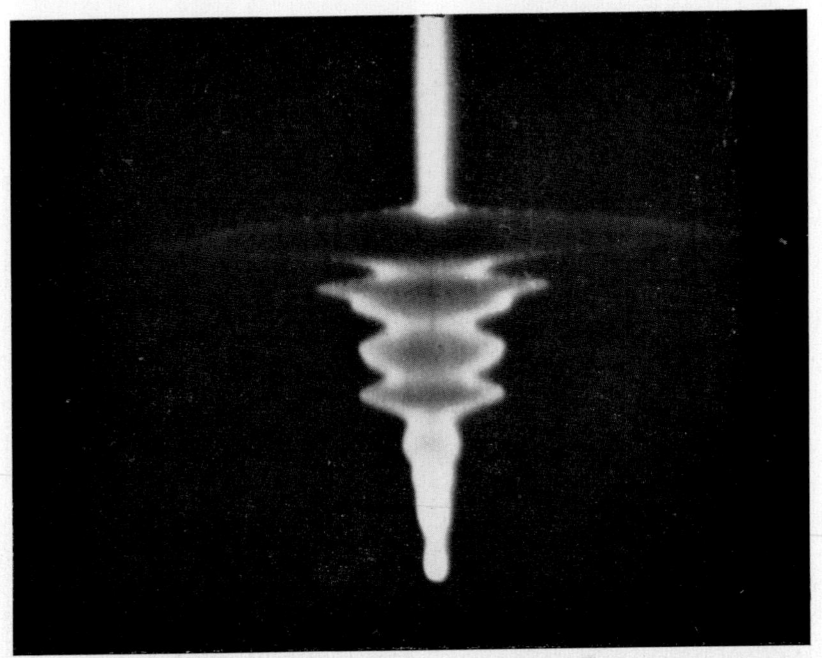

Fig. 5.25 A typical sea-bed echo as seen on an expanded CRT display. The time base is vertical and is 15 ms long (range increasing downwards)

speed of the pen is extremely high and its motion is triggered by the sea-bed echo, the signals being delayed by 20 ms before being applied to the recorder. A typical record is seen in Fig. 5.26(a) (which was taken at the same time as Fig. 5.26(b)). The thin black line at the foot of this record is the initial rise of the sea-bed echo. The annulus of range above the sea-bed echo which is examined on this expanded record, is only 4 echo-fathoms high. The individual fish are clearly distinguished. (Time marks, every minute, are also seen at the top of the record.)

These two types of records give distorted views of the distribution of fish directly under the trawler as it moves along.

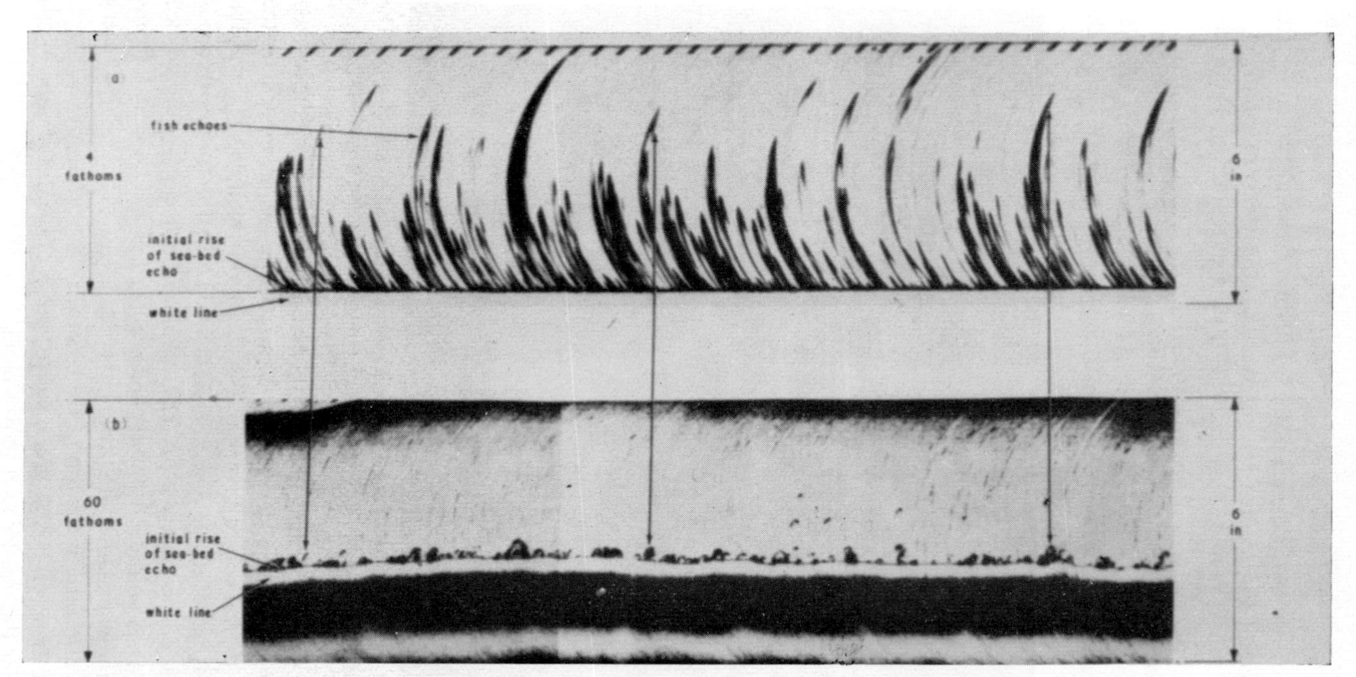

Fig. 5.26 Typical charts obtained with a recording scale-expander, compared with corresponding normal echo-sounder charts taken at same time.[24] Numerous fish-echoes are seen above the sea-bed echo: some points corresponding in time are indicated:

(a) scale-expander chart } depth 100 fathoms; fish, cod and haddock,
(b) normal chart } catching rate: 30 baskets/hour

It is clear that to interpret observations from the CRT scale-expander and from these recorders, detailed information regarding the acoustic back-scattering properties of fish is required.

5.17 Back-scattering polar diagrams of a fish

The lengths of some of the fish caught by the fishing industry at sea are seen in Table 5.3.

Table 5.3 *Lengths of some commercially important fish*[25]

Name	Overall length (L) cm	In wavelengths at 30 kHz†
Cod	45–120	9–24
Hake	30–90	6–18
Haddock	30–60	6–12
Whiting	20–40	4–8
Herring	15–30	3–6
Pilchard	12–25	2·4–5
Sprat	8–15	1·6–3

† Most instruments use this frequency (wavelength in sea water = 5 cm).

Using the scale-model technique, the back-scattering polar diagrams of a fish have been determined at 1·48 MHz, 625 kHz and 360 kHz over a wide range of fish lengths[25].

Typical polar diagrams are seen in Fig. 5.27 for model fish which correspond to small, medium and large fish at 30 kHz. The model fish usually consisted of sticklebacks which have a similar shape and structure to that of full-size fish.

Each fish was mounted on the needle support (see Section 5.5) in two positions.

(i) In the normal swimming position in which readings were taken of the signals back-scattered by the fish whilst it was rotated about a vertical axis (peak amplitude irrespective of pulse shape). This plane of reference (termed 'horizontal plane') included the head (0°), tail (180°) and the two broadside aspects (90° and 270°) and is seen on the left in Fig. 5.27.

(ii) Readings were taken with the fish mounted tail uppermost whilst it was rotated about its longitudinal axis. This plane of reference (termed 'vertical plane' included the dorsal (0°), ventral (180°) and the two broadside aspects (90° and 270°) and is shown on the right of Fig. 5.27.

In the horizontal plane, the polar diagrams are characterized by a large maximum on each side, somewhat towards the tail in the case of larger fish

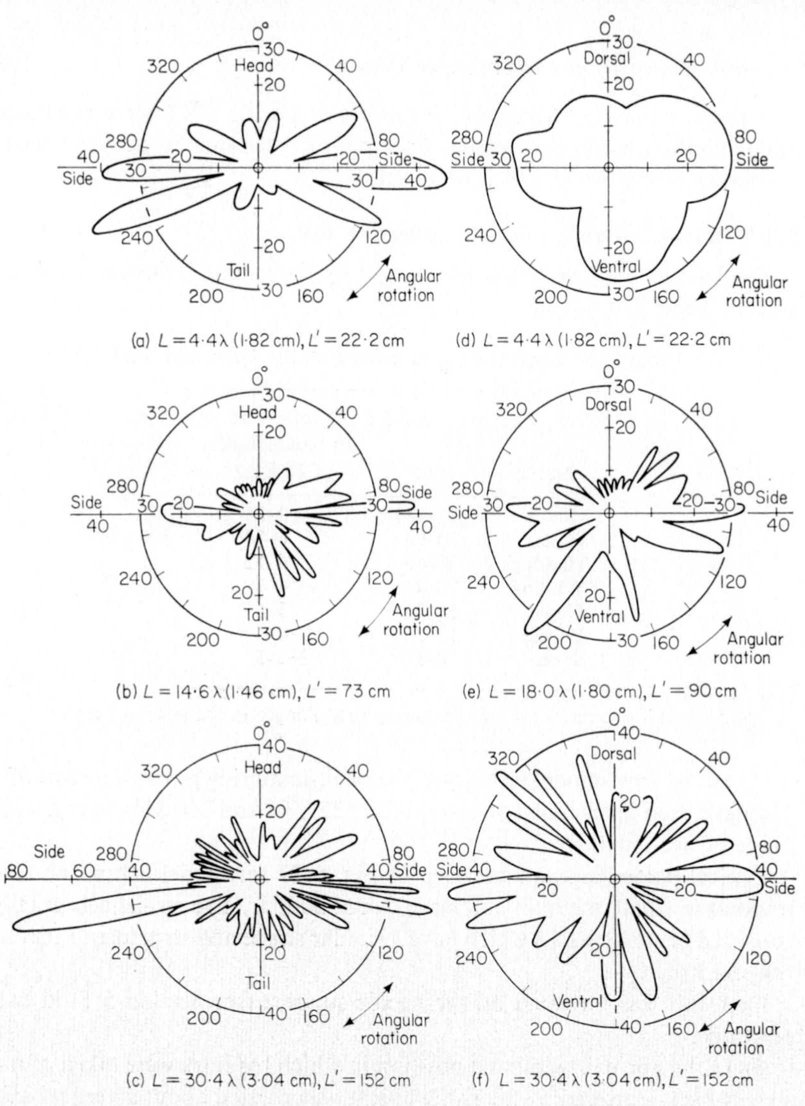

(a) $L = 4{\cdot}4\lambda$ (1·82 cm), $L' = 22{\cdot}2$ cm
(d) $L = 4{\cdot}4\lambda$ (1·82 cm), $L' = 22{\cdot}2$ cm
(b) $L = 14{\cdot}6\lambda$ (1·46 cm), $L' = 73$ cm
(e) $L = 18{\cdot}0\lambda$ (1·80 cm), $L' = 90$ cm
(c) $L = 30{\cdot}4\lambda$ (3·04 cm), $L' = 152$ cm
(f) $L = 30{\cdot}4\lambda$ (3·04 cm), $L' = 152$ cm

Fig. 5.27 Back-scattering polar diagrams of fish (sticklebacks)[25]. The readings were taken under the following conditions:

Plane of observation: (a), (b) and (c), horizontal plane; (d), (e) and (f), vertical plane (perpendicular to longitudinal axis of fish).

Frequency: (a) and (d), 360 kHz; (b), (c), (e) and (f), 1·48 MHz.

Radial scale: amplitude (mV): 20 mV in (a) and (d) is equivalent to an acoustic cross-section of $1{\cdot}5 \times 10^{-2}$ cm^2 at 360 kHz (or $8{\cdot}8 \times 10^{-4}$ cm^2 at 1·48 MHz); 20 mV in (b), (c), (e) and (f) is equivalent to an acoustic cross-section of $1{\cdot}73 \times 10^{-1}$ cm^2 at 1·48 MHz.

L = actual length of fish, L' = equivalent full-size length of fish at 30 kHz, λ = wavelength in water

(e.g., 260° and 100° in Fig. 5.27(c)). This is termed the 'maximum side signal' and those in tail and head aspects are usually smaller by comparison. In the vertical plane, the signals rise to a maximum in side aspect but the effect is not so pronounced.

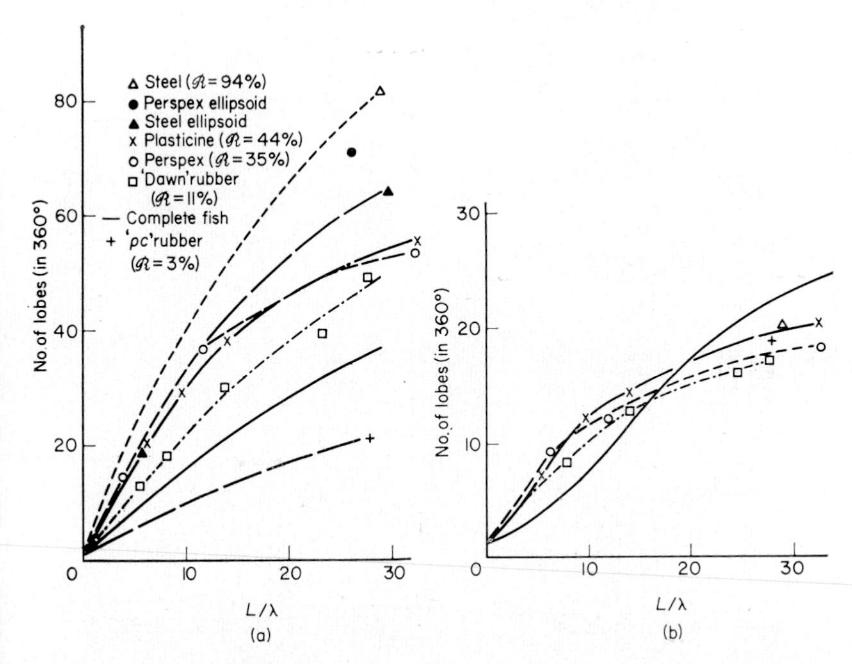

Fig. 5.28 The numbers of lobes in the back-scattering polar diagrams of models of the standard fish body in various materials, plotted against overall length of fish (L)[34]. (a) horizontal plane, (b) vertical plane. For comparison similar graphs for actual fish are given. ($\mathscr{R}$ is the amplitude coefficient of reflectivity at a single plane interface immersed in tap water)

The number of lobes in a complete polar diagram increases for larger fish as seen in Fig. 5.28 (full lines). (Other details in this figure are considered in Section 5.22.4.)

This complex system of lobes exists in three dimensions around the fish. At a fish length of between 14 λ and 18 λ (Fig. 5.27(b) and (e)), the polar diagrams are less clearly defined.

For absolute measurements of back-scattering cross-section, a standard target in the form of a steel 'disc' was substituted for the fish. For practical purposes this acted as a single interface (between water and steel) in the

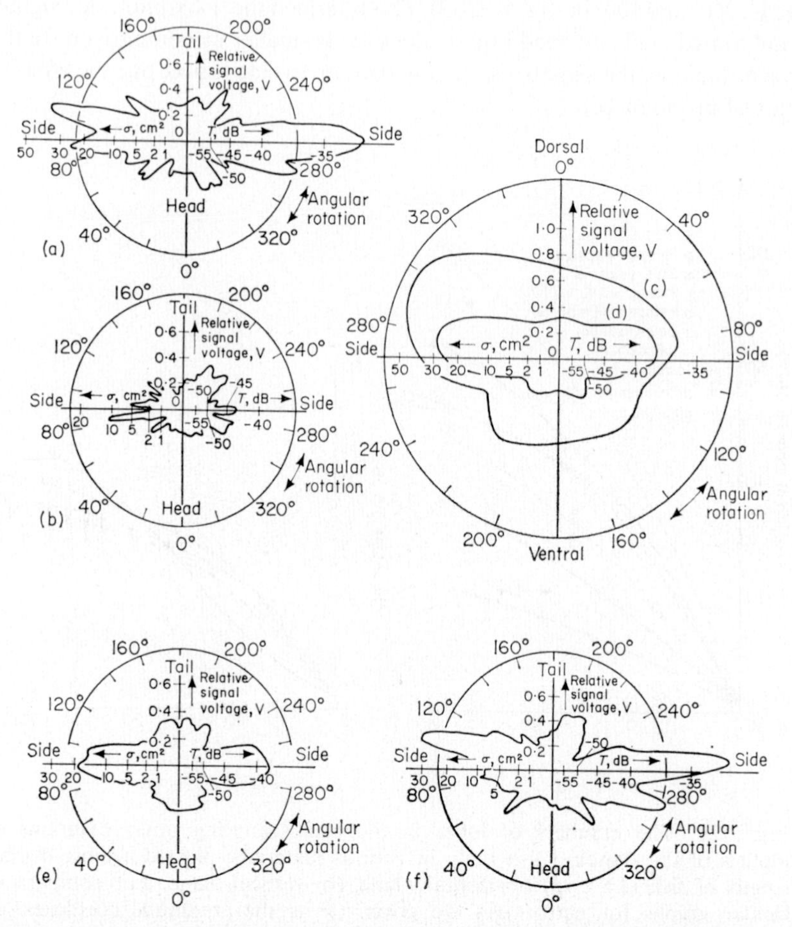

Fig. 5.29 Acoustic back-scattering polar diagrams of a perch of length 20 cm (4 λ) at 30 kHz (after Jones and Pearce). The radial coordinate is given in terms of relative signal voltage, back-scattering cross-section (σ) and target strength (T) compared with a perfectly reflecting sphere of radius 2 m[26].

(a) Swimbladder full of air, horizontal plane
(b) Swimbladder emptied of air, horizontal plane
(c) Swimbladder full of air, vertical plane
(d) Swimbladder emptied of air, vertical plane
(e) Artificial swimbladder alone, horizontal plane
(f) Fish with artificial swimbladder, horizontal plane

form of a plane disc, so that its acoustic back-scattering cross section in broadside aspect could be calculated at the three frequencies.

These results from the scale model have been confirmed on full size fish by Jones and Pearce[26] using perch at 30 kHz. There is a tendency for the swimbladder to collapse when the fish is handled. To get over this difficulty, for each fish they made an artificial swimbladder of Onazote (which is a form of expanded plastic containing air bubbles) and implanted it in the fish. Fig. 5.29(a) shows the polar diagram in the horizontal plane for an unadulterated fish of length 20 cm while (b) gives the same polar diagram when the swimbladder was filled with water. In broadside aspect, the signal voltage has declined by about 50 per cent. Similar polar diagrams taken in the vertical plane, are seen in (c) and (d) and again the decline is about 50 per cent voltage in dorsal and ventral aspects. The artificial swimbladder was 5 cm long and 1·2 cm diameter and, alone, gave the polar diagram (e), showing maxima in broadside aspect approaching those seen in diagram (a). When this was embedded in the fish inside its swimbladder, the polar diagram (f) was obtained. Comparison with (a) shows that the results are substantially the same, from which it may be inferred that the Onazote swimbladder has similar acoustic properties to the actual swimbladder when air-filled.

The similarity between Fig. 5.29(a) taken at full scale and Fig. 5.27(a) at model scale for fish of approximately the same length (4 λ) is to be noted, also between Fig. 5.29(c) and 5.27(d). (N.B. Fig. 5.29(a) is head-down whilst Fig. 5.27(a) is head-up.)

5.18 Effect of fish length on acoustic cross-section

The effect of fish length on the energy scattered in the main directions may be drawn graphically (Fig. 5.30). The 'actual broadside signal' was the mean amplitude in true side aspect (i.e., exactly at 90° and 270° in the horizontal plane). Only the 'maximum side signal' could be determined without averaging, as this was clearly defined. Since there appeared to be little difference between the dorsal and ventral signals, the mean value was taken.

The signals in these main directions are seen to have regularly recurring maxima and minima. The peaks of 'maximum side' amplitude are always larger than those for 'dorsal' and the peaks of 'actual side' amplitude lie between these two. Clearly interference occurs between the signals from two (or more) parts within the fish.

Fig. 5.27(b) ($L = 14{\cdot}6\ \lambda$) is seen to correspond to a minimum in the maximum side signal; also Fig. 5.27(e) ($L = 18\ \lambda$) to a minimum of the dorsal signal. It appears that the back-scattering polar diagrams become distorted

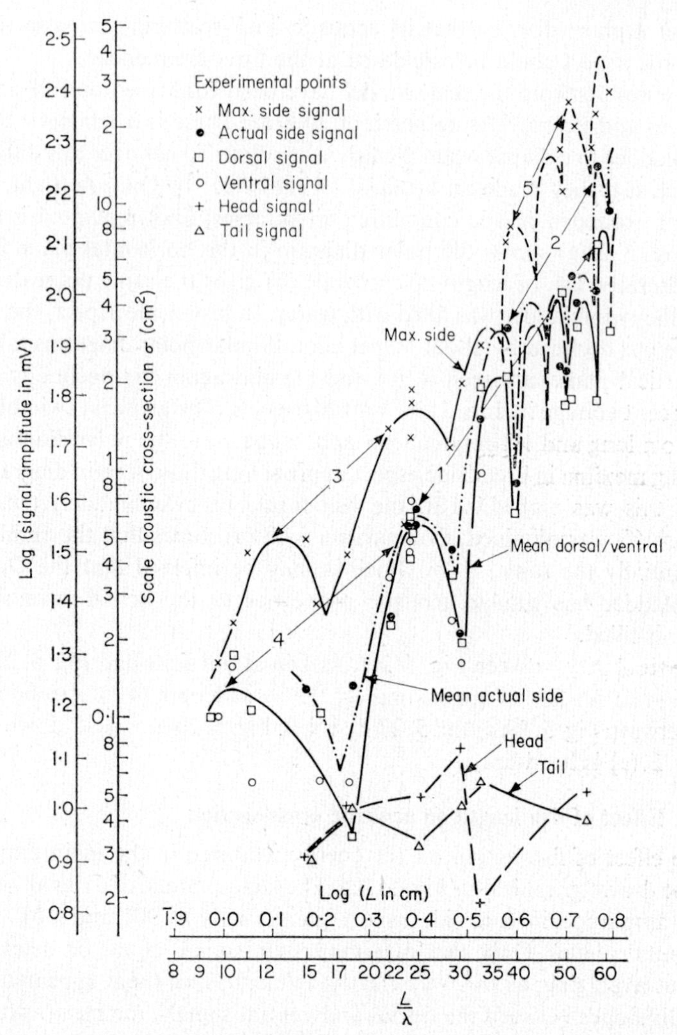

Fig. 5.30 Acoustic cross-sections of sticklebacks and guppies in main aspects, measured at 1·48 MHz[25]. (L = overall length of fish, λ = wavelength in water.) The numbered straight lines show the general trends of the cross sections at various points:

Line	Slope
1 (dorsal)	2·9
2 (actual side)	3·1
3 (maximum side)	1·8
4 (dorsal)	1·6
5 (maximum side)	4·1

at these minima, the general effect being to make the polar diagrams more omnidirectional.

As has been observed (in Section 5.6), a cylinder would have an acoustic cross-section proportional to the cube of length if its radius a' were also proportional to length (as occurs in a fish of constant shape). On examining Fig. 5.30, if interference effects are ignored, the slopes of the general trends between $L=24\ \lambda$ and $L=60\ \lambda$ (lines 1 and 2) are found to be 2·9 (dorsal) and 3·1 (actual side), thus confirming that some part of the fish (swimbladder or backbone?) acts in the same way as a cylinder and that its signal predominates for large fish in these aspects.

The simple geometrical body which most closely approximates to the fleshy body of a fish is an ellipsoid which has an acoustic cross-section proportional to the square of length, if constant shape is maintained. The slopes of line 3 and 4 are 1·8 (maximum side) and 1·6 (dorsal) respectively and suggest that the signal from the body of fish flesh may predominate for small fish (in the range $L=10\ \lambda$ to $34\ \lambda$).

These experimental results accord with theory. At constant frequency, when the size of the fish is increased, the fleshy body will pass from the Rayleigh scattering region into the geometrical region *before* the swimbladder or backbone, since the latter structures are smaller than the body. Thus the initial general trend for small fish should approximate to an ellipsoid (i.e., proportional to the square of fish length). However, for larger fish the signals from the cylindrical bladder and backbone (increasing according to a cube law) will exceed that from the body.

It is interesting to note that for large fish (greater than $35\ \lambda$) the general trend of the maximum cross section side (line 5) has a slope of 4·1. This suggests that in this aspect, the fish acts like a plane target in which the shape is kept constant when the length is increased, which would have a cross section increasing according to the fourth power of length.

5.19 Frequency response of a fish

At very low frequencies and for a constant fish size, the fish will be in the Rayleigh scattering region and the cross section will be proportional to the fourth power of frequency (unless a resonance occurs in the air-filled swimbladder). As the frequency is raised, the fleshy body (ellipsoid) will pass into the geometrical region before the swimbladder and backbone, so that initially the body signal will predominate independent of frequency. At higher frequencies, the signals from swimbladder and backbone (cylinders) will predominate as their cross sections are proportional to frequency. If a plane reflecting area were involved (e.g., in side aspect?), this would finally predominate at high frequencies as its cross section is proportional to the square of frequency.

These results are supported by Hashimoto's calculations for dorsal aspect, based on the cylinder, also by the suggestion of Hashimoto and Maniwa of a formula for side aspect based on a plane reflecting area. (Both

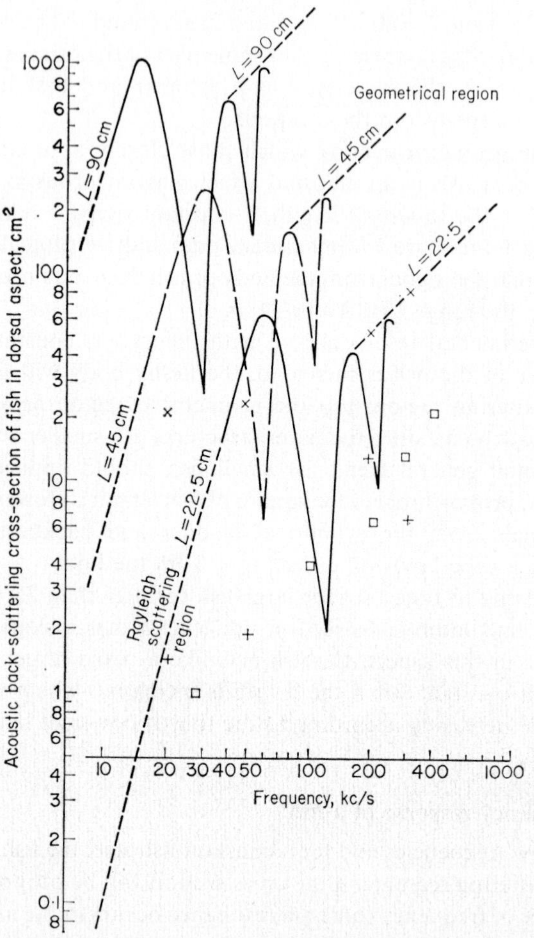

Fig. 5.31 Approximate frequency responses of fish of lengths 90 cm, 45 cm and 22·5 cm determined by scale-model measurements in the geometrical region. At high frequencies there are many maxima, so the general trend of the maxima is shown as a dotted line on each graph[25].

These results are compared with those published by Hashimoto and Maniwa for a fish of length 15 cm (□) and by Shishkova for fish of lengths 45 cm (×) and 22·5 cm (+)

proposals are valid only when the frequency is high and refer to general trends, ignoring interference effects.)

The detailed shape of the frequency response of a fish may be calculated from Fig. 5.30 using the correct scaling factor for each point. Three examples are given in Fig. 5.31 for fish of lengths 90 cm, 45 cm and 22·5 cm and are compared with some results due to Hashimoto and Maniwa[27, 28], also Shishkova[29], which are found to be in general agreement.

These frequency responses are also confirmed by Cushing and Richardson[30] who operated three echo-sounders simultaneously at 10 kHz, 14 kHz and 30 kHz on a ship at sea. After careful statistical analysis and corrections to eliminate the effects of the different beam-angles and transmitted powers, they found large differences in the amplitudes at the three frequencies for fish of various sizes. The signal ratios measured by Cushing and Richardson appear to accord with Fig. 5.31, particularly for the smaller fish in the Rayleigh scattering region.

More recently, McCartney[31, 32] has obtained some results on larger fish at very low frequencies which suggest the existence of resonance of the swimbladder.

5.20 Structure of a fish

The parts of a fish which mainly contribute to the overall echo are seen in Fig. 5.32[33]. Apart from the body consisting of fish flesh, other important structures are the swimbladder and the vertebral column (or 'backbone'). The swimbladder is a cigar-shaped air-filled envelope situated just under the backbone. The gut consists mainly of tissue of much the same density as the

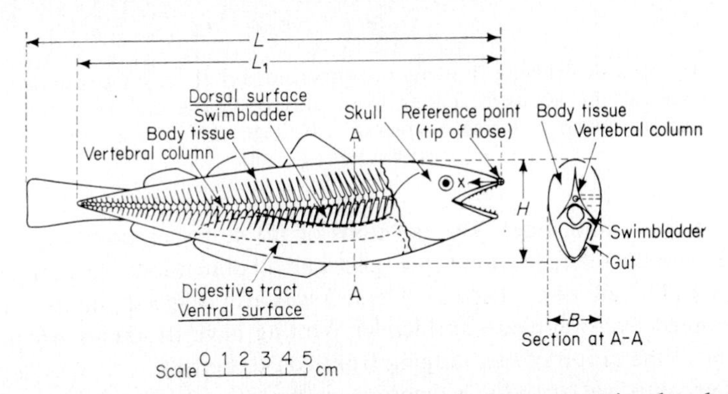

Fig. 5.32 Parts of a fish (e.g., whiting) contributing to its acoustic echo, showing nomenclature concerning dimensions and the coordinate (X) which was used to fix position in the fish. L=overall length of fish; H=maximum height of body (less fins); B=maximum breadth of body (less fins)[33]

flesh and is not expected to contain much gas. The fins are ignored as they do not reflect appreciable amounts of sound since they are so thin as to be acoustically transparent.

To calculate the magnitudes of the various contributions to the overall echo from different parts of a fish, the structure must be examined in some detail. Several freshly-caught whiting were dissected, some transversely and others longitudinally, care being taken to observe the exact positions of the

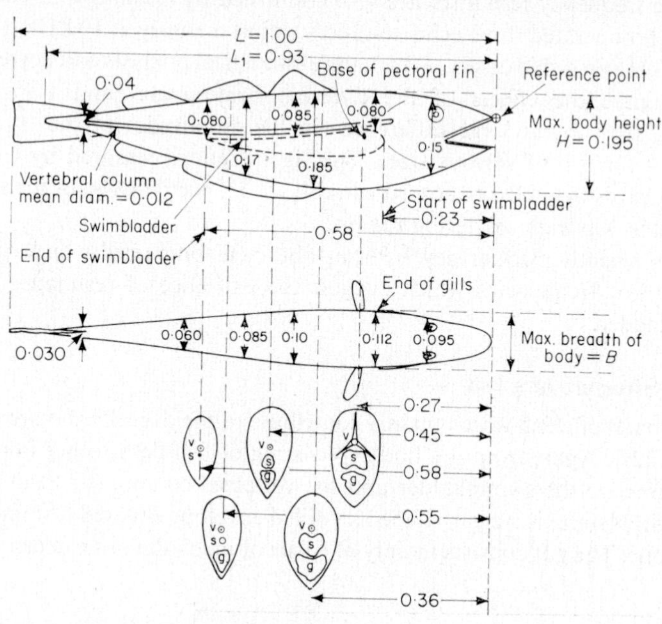

Fig. 5.33 A scale drawing showing the dimensions of the average fish on which calculations of the acoustic echoes from different parts can be based[33]. All dimensions are given in terms of overall fish length L[33]. v=vertebral column; s=swimbladder; g=gut

ends of the swimbladder. The positions of the sectional views were measured from the reference point (the nose) and all dimensions were related to the overall length of the fish (L). The fish was also weighed and its volume determined by Archimedes' principle. Whiting have structures which are typical of the group of fish ranging from cod to herring.

After statistical analysis, a diagram giving the mean dimensions of the fish (the 'standard fish') was drawn (Fig. 5.33) which gives the average measurements relating to the positions and shapes of the swimbladder and the vertebral column, as well as the thicknesses of tissue between these structures and the sea water. In the normal swimming position, the vertebral column curves downwards a little in the middle.

The volume of each swimbladder was calculated from the measurements by dividing into a number of simple figures amenable to calculation. The mean volume of the swimbladder was 4·1 per cent of the total volume of the fish. The mean volume of the whole fish was $8{\cdot}3 \times 10^{-3}\ L^3$ cm³.

5.21 Approximations to parts of a fish

5.21.1 Body

The body of the fish (less fins) approximates to an ellipsoid of dimensions: $L_1 = 0{\cdot}93L$, $H = 0{\cdot}195L$ and $B = 0{\cdot}112L$. This approximation will be reasonably accurate for dorsal, lateral and head aspects, but less accurate for tail aspect where the shape of the fish differs substantially from that of the ellipsoid. The mean acoustic impedance of wet fish flesh has been found to be $1{\cdot}6 \times 10^5$ cgs units[11] giving an amplitude reflectivity ($\mathscr{R}$) of 4·4 per cent in fresh water or 1·9 per cent in salt water, considering the front interface only.

5.21.2 Swimbladder

Fig. 5.34(a) gives an enlarged view of the region of the fish which is important for echo sounding, where the sound waves enter the fish through the dorsal surface.

An approximation can be made to the average swimbladder (Fig. 5.34(b)) consisting of a cone ABC, a cylinder BDEC and a hemisphere DFE. As the swimbladder contains gas, it is perfectly reflecting to sound. Although it is possible to calculate the signals reflected from these bodies, the method is unwieldy and it is simpler, although less accurate, to use as an approximation in the geometrical region, the cylinder GDEH having a length $0{\cdot}24L$ and a radius $0{\cdot}0245L$. The swimbladder, being curved, will distribute the reflected sound somewhat more omnidirectionally than this cylinder.

5.21.3 Vertebral column

The core of the vertebral column is a cylinder of length about $0{\cdot}65L$ having a mean diameter of about $0{\cdot}012L$. The reflectivity of wet fish bone[11] has been found to be 26 per cent (considering reflection at the front interface only). It may be necessary to make a correction for the curvature of the vertebral column when the fish is many wavelengths long. The signals reflected from the bony spines of the backbone are generally much smaller than from the core.

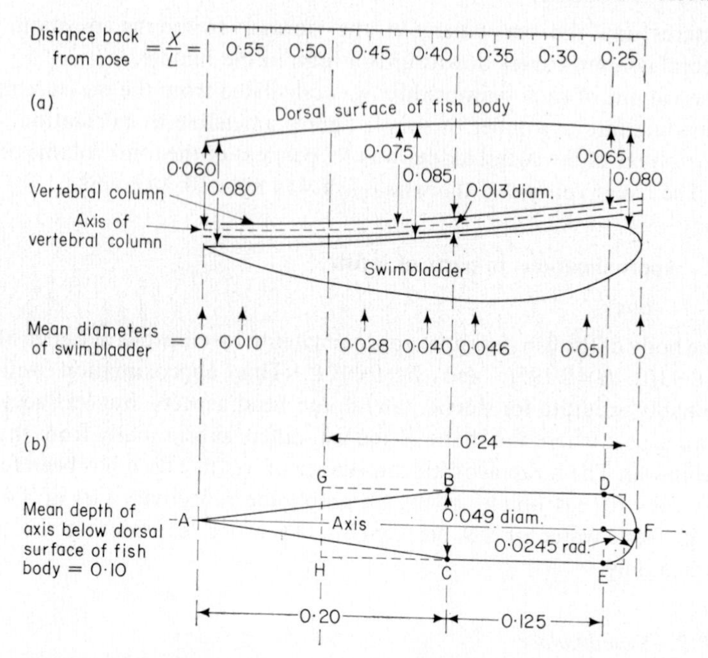

Fig. 5.34 (a) Enlarged view of mean swimbladder, showing dimensions relevant to echo sounding. (b) Two approximations to the shape of the swimbladder, using simple geometrical bodies ($\mathscr{R} = -100$ per cent). All dimensions are expressed in terms of L, the overall length of the fish[33]

If necessary, corrections could be also made for deviations from the mean shapes of all the structures within a fish.

5.22 Acoustic properties of models of fish body and ellipsoids

An investigation of the back-scattering properties of models of a fish *body* in various materials has been carried out (omitting the swimbladder and the bones) and the results have been compared with those of ellipsoids having the equivalent dimensions (given in Section 5.21.1)[34]. Since a calculation of the signal back-scattered by the actual shape of a fish is intractable, the scale-model method of measurement was used.

5.22.1 *Fish body*

The models, in a range of sizes, were constructed of various materials, some translucent and some opaque to sound. Each model was rotated and readings were taken at 1·48 MHz as described for the small fish in Section 5.17). A signal of 20 mV was equivalent to an acoustic cross-section of

$1{\cdot}73 \times 10^{-1}$ cm^2 and the scaling factor was 50/1. (The full-size acoustic cross-section at 30 kHz may be obtained by multiplying the scale cross-section by the square of this factor.)

The materials used for the models (and their amplitude reflectivities $\mathscr{R}$ at a plane interface, immersed in tap water) were: steel (94 per cent), Plasticine (44 per cent), Perspex (35 per cent), 'Dawn' rubber (11 per cent) and 'ρc' rubber (3 per cent). As steel is almost opaque to sound, the results using this material indicate substantially the effect of the front surface only. This also applies to Plasticine in which the acoustic absorption is so high (19 dB cm^{-1}) that the echo from the rear interface is greatly reduced compared with that from the front surface.

Perspex and the two forms of rubber are translucent to sound and have low values of acoustic absorption so that the combined effect of the two reflecting surfaces (front and rear) is shown. By plotting the results for these substances covering a range of acoustic reflectivities, the amount of sound energy scattered from a body of fish flesh (of reflectivity $\mathscr{R}=4{\cdot}6$ per cent) is obtained by interpolation.

Typical polar diagrams of back-scattering are seen in Figs. 5.35–39. Readings in the 'horizontal plane' were obtained while the target was rotated about the dorso-ventral axis. In the case of the 'vertical plane', the target was rotated about its longitudinal axis. The form of the diagram for the body of a fish (without swimbladder and bones) would lie between Figs. 5.38 and 5.39.

Since Figs. 5.35–39 give the results for a wide variety of materials of reflectivities from 3 per cent to 94 per cent, by interpolation they would also be useful in estimating the polar diagrams and back-scattering cross-sections of a body of similar shape in almost any homogeneous material (for example, a towed body) for it is the length in wavelengths that determines its acoustic performance.

5.22.2 *Ellipsoids*

Although an ellipsoid of dimensions L_1, H and B is a simple approximation to a fish (see Section 5.21.1), this is difficult to manufacture, so ellipsoids of revolution having a mean value between H and B were used instead. The back-scattering polar diagrams of these targets are seen in Fig. 5.40. The acoustic back-scattering cross sections of the equivalent ellipsoid are $\sigma=\pi(L_1 B\mathscr{R}/2H)^2$ for dorsal aspect and $\pi(L_1 H\mathscr{R}/2B)^2$ in side aspect (for the front surface only). Thus the echo-voltages will be proportional to B/H and H/B respectively and the readings for the ellipsoids of revolution may be corrected to indicate the signals expected from the ellipsoidal approximation to the fish body.

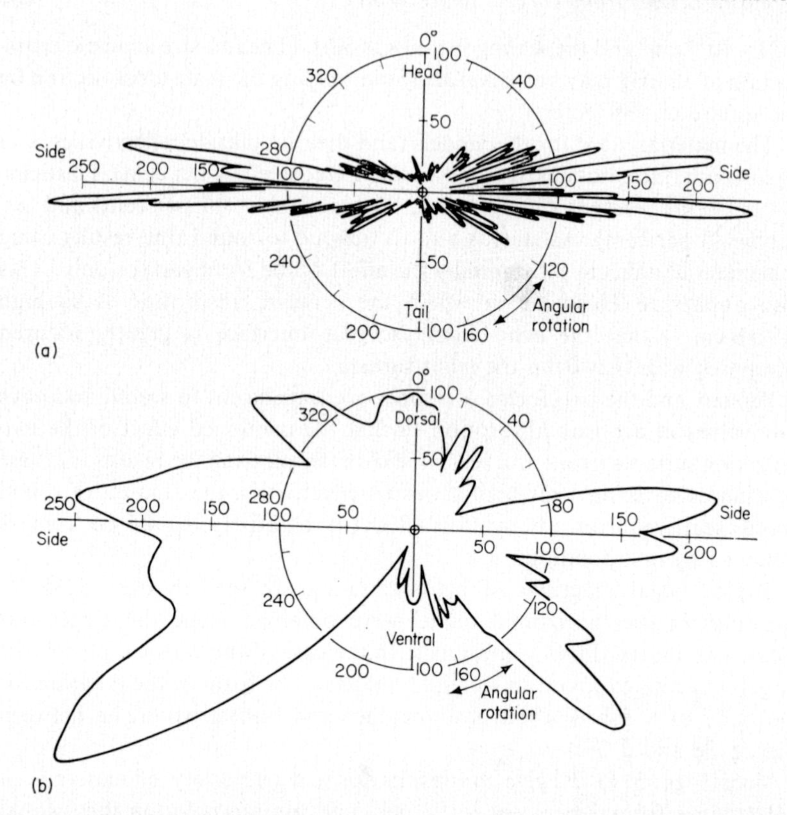

Fig. 5.35 Back-scattering polar diagrams of a model of the standard fish body in steel[34]. (a) horizontal plane; (b) vertical plane. Radial scales: amplitude (mV). $L_1 = 2{\cdot}66$ cm, $H = 0{\cdot}555$ cm, $B = 0{\cdot}326$ cm, $L = 2{\cdot}86$ cm ($28{\cdot}6\ \lambda$), $L' = 143$ cm

5.22.3 *Shapes of polar diagrams*

As is to be expected, the directivity of the pattern increases with target size, and, at the same time, the angular separation of the lobes becomes smaller. When the targets are made of materials translucent to sound, the effect of the rear surface is to scatter the energy more uniformly. This effect increases if the material is more translucent.

For a constant size of target, the angular separation of the lobes increases as the reflectivity of the material falls. In general, the patterns in the horizontal plane for the ellipsoids (Fig. 5.40) are less directional than those for the fish models (Figs. 5.35–39). This effect is attributed to the difference of shape, the sides of the models being flatter.

On comparing these polar diagrams for the models of the body (Figs.

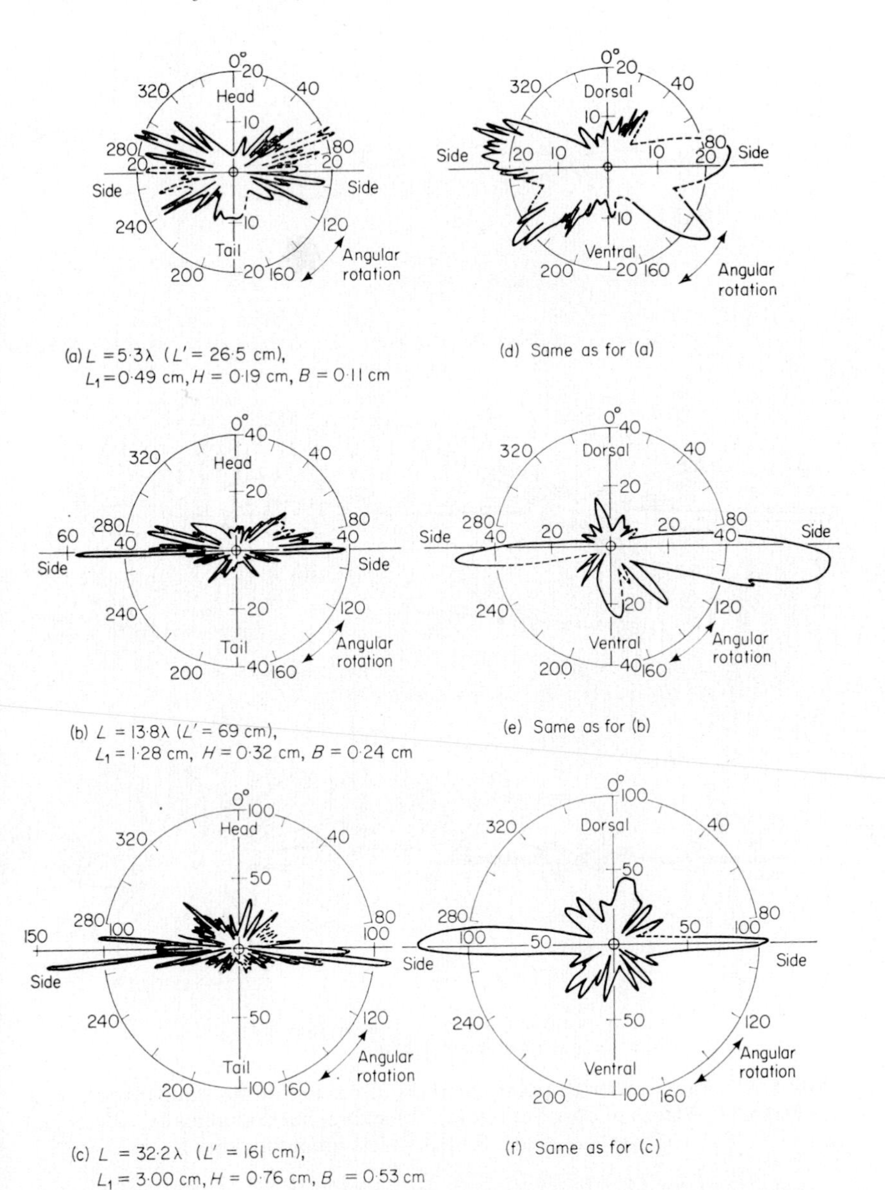

Fig. 5.36 Back-scattering polar diagrams of models of the standard fish body in Plasticine[34]. Planes of observation: (a), (b) and (c) horizontal plane; (d), (e) and (f) vertical plane. Radial scales: amplitude (mV)

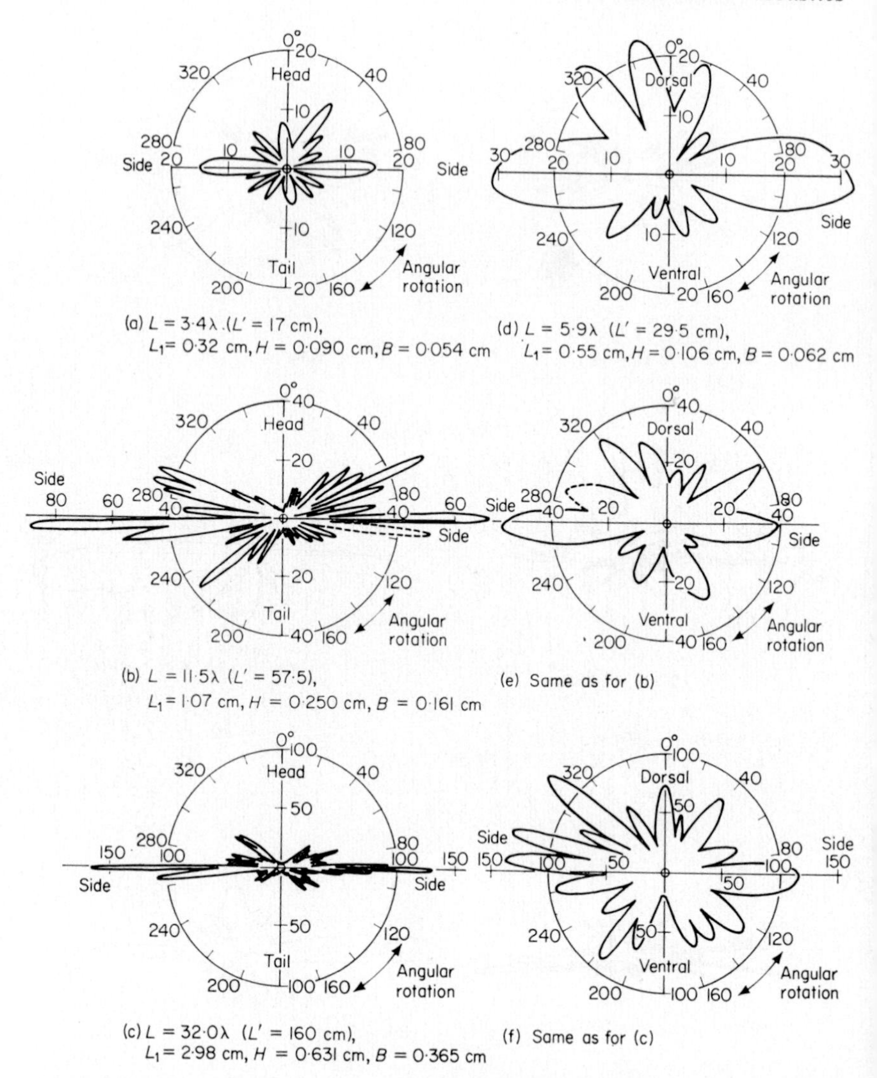

Fig. 5.37 Back-scattering polar diagrams of models of the standard fish body in Perspex[34]. Planes of observation: (a), (b) and (c) horizontal plane; (d), (e) and (f) vertical plane. Radial scales: amplitude (mV)

5.35–39) with corresponding patterns for the actual fish (Fig. 5.27), it is seen that the fish scatter sound more uniformly than do the models (although the difference is much less in Fig. 5.39 in the case of 'ρc' rubber). This is attributed to the presence of other scatterers within the fish body

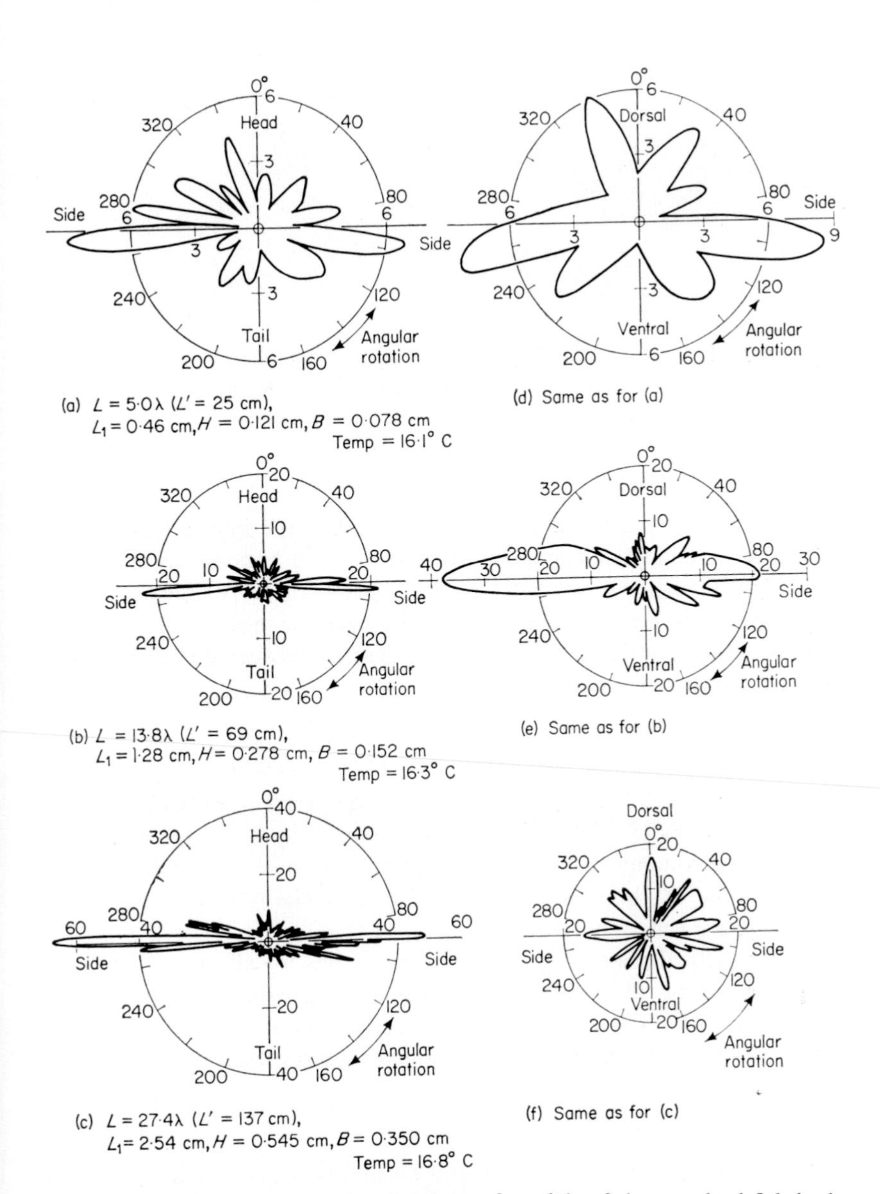

Fig. 5.38 Back-scattering polar diagrams of models of the standard fish body in 'Dawn' rubber[34]. Planes of observation: (a), (b) and (c) horizontal plane; (d), (e) and (f) vertical plane. Radial scales: amplitude (mV)

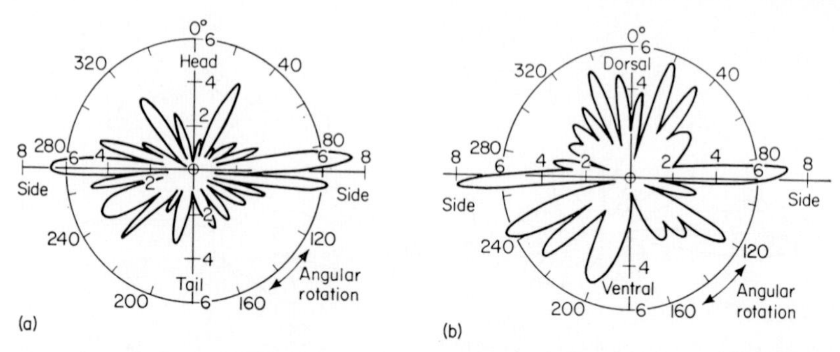

Fig. 5.39 Back-scattering polar diagrams of a model of the standard fish body in 'ρc' rubber[34]. (a) horizontal plane; (b) vertical plane. Radial scales: amplitude (mV). $L_1 = 2{\cdot}56$ cm, $H = 0{\cdot}535$ cm, $B = 0{\cdot}36$ cm, $L = 2{\cdot}75$ cm ($27{\cdot}5\ \lambda$), $L' = 138$ cm. (Temperature $= 17{\cdot}9$°C)

(e.g., swimbladder and vertebral column). Also, the interference minima which occurred at fish lengths between 14 λ and 18 λ and which are accompanied by distortion of the polar diagrams (see Fig. 5.27(b) and (e)), are not observed in the measurements on models, since the models have no swimbladder or vertebral column, so that interference between the various signal contributions is eliminated.

5.22.4 Comparison of numbers of lobes in polar diagrams

The numbers of lobes in the complete polar diagrams of back-scattering (horizontal plane) for fish models and fish (see in Fig. 5.28(a)) is empirically found to be roughly proportional to length (L), to increase with $\mathscr{R}$ and to have a value of about 80 $\mathscr{R}^{1/3}$ when $L = 30\ \lambda$. The numbers of lobes for ellipsoids are generally similar.

In the vertical plane (Fig. 5.28(b)), the graphs for all the models are in close agreement, the number of lobes increasing with length, with about 20 lobes at $L = 30\ \lambda$. The graph for fish is somewhat different and this is again attributed to the presence of other structures within the fish.

5.22.5 Comparison of signal levels

The maximum side signal follows a linear law of amplitude against fish length, since this method of measurement ensures that the signals from front and rear surfaces add to give a maximum value. (As the target was rotated, the phase difference between the echoes from front and rear surfaces was changed.)

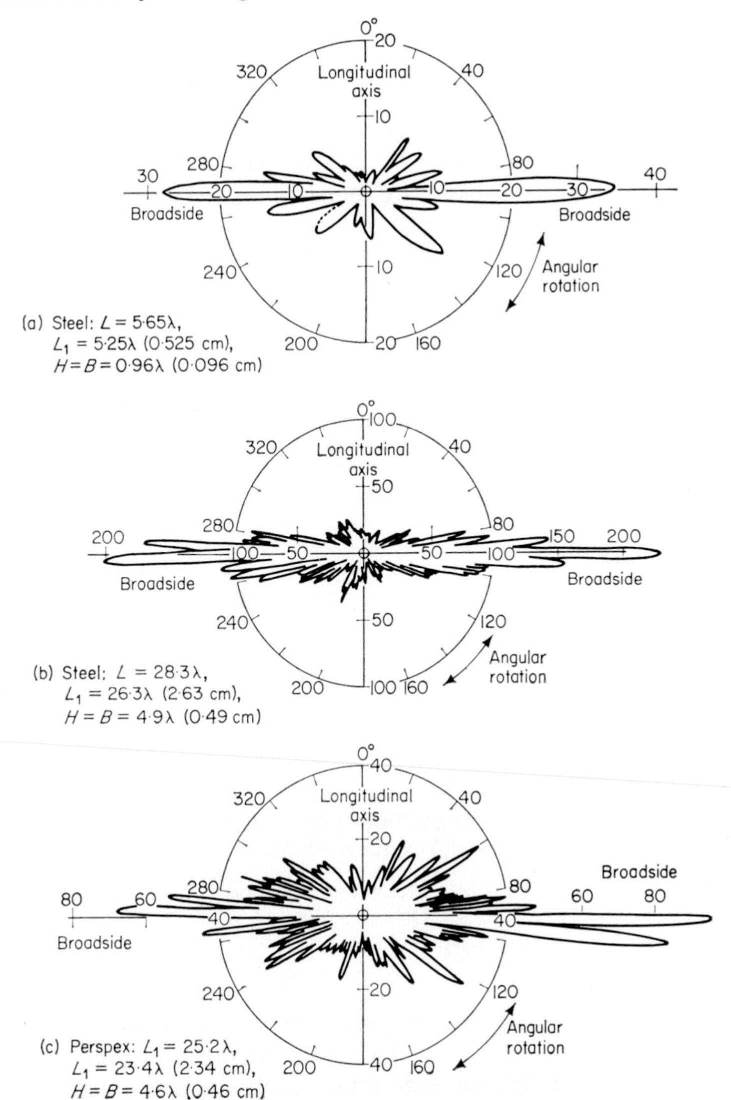

Fig. 5.40 Back-scattering polar diagrams of ellipsoids of revolution made of steel or Perspex, taken in the horizontal plane[34]. Radial scales: amplitude (mV)

The approximate mean slopes (versus length) for dorsal and ventral aspects follows power laws of 0·8 and 0·55 respectively.

In an approximation, the first two significant reflections may be considered, that is the first reflection from the front surface and the first from the rear surface (and further inter-reflections may be ignored). Also assuming that the acoustic cross-section of the rear surface approximates to that of the front surface, a rough value of total reflectivity $\mathscr{R}_T$ may be obtained.

In Fig. 5.41, the signals received from the models in various aspects were corrected to an overall length of $L=2{\cdot}6$ cm. The signals for the ellipsoids corrected for length and shape, are found to be similar in level to those

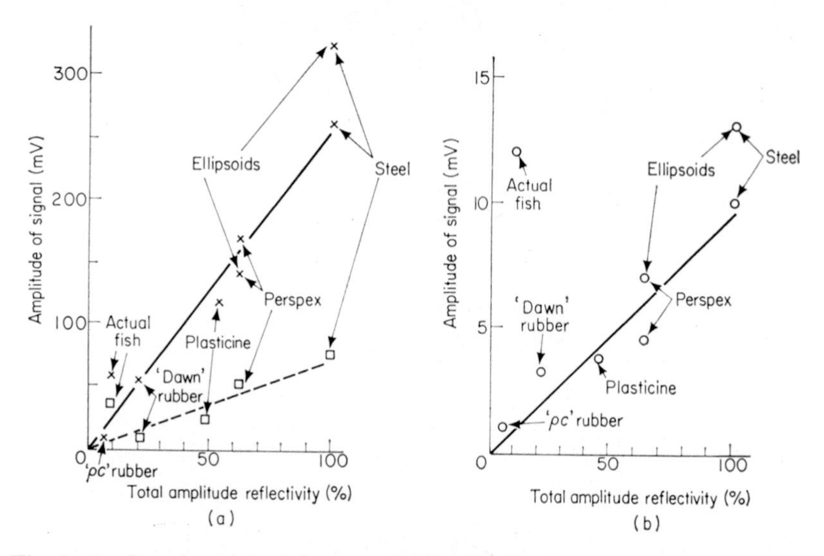

Fig. 5.41 Signals received from model fish bodies in various materials, compared with those from ellipsoids and from small fish. In all cases, the equivalent overall length is 2·6 cm[34]

(a) × Maximum side aspect
□ Dorsal aspect
(b) ○ Head aspect

from the models, but those from the actual fish are seen to be much higher than calculated on the basis of the fleshy body alone (showing that account must also be taken of the swimbladder and vertebral column). The straightness of the lines in Fig. 5.41 substantially confirms the accuracy of the assumptions.

These measurements show that the equivalent ellipsoid having the same acoustic properties as fish flesh ($\mathscr{R}=4{\cdot}6$ per cent) (for example soft rubber) would serve as a satisfactory approximation to the fleshy body of a fish and would give similar echo-levels and polar diagrams (except for tail aspect where accuracy is impaired). This can be used as a basis for calculations.

5.23 Calculations of echo contributions from parts of a fish

A table of formulae for the acoustic cross-sections of various parts of a fish may now be drawn up (Table 5.4). Over 99 per cent of the incident energy passes through the tissue so that the values of $\mathscr{R}$ for the swimbladder and bone immersed in water can be taken, although these structures are really surrounded by fish flesh. These formulae are based on Table 5.1 and the approximations of Section 5.21.

In regard to long fish at high frequencies, it is clear that the fish acts as a plane reflecting area in broadside aspect since under these conditions the general trend of the acoustic cross-section in Fig. 5.30 (line 5) accords with a fourth-power law of fish length. The exact cause is at present obscure. It may be that the regular arrangement of the upper spines of the backbone in a plane lattice causes the effect. In this case the average effective value of $\mathscr{R}$ per unit area might be as high as that of fish flesh. The latter is taken to give a guide to the upper limit of signal to be expected from this effect.

5.23.1 Cylindrical core of fish backbone

A cylinder of damp fish bone (having a reflectivity $\mathscr{R}=26$ per cent) would be expected to behave in the Rayleigh scattering region in a similar manner to that of Nylon or Perspex seen in Fig. 5.13 but somewhat lower on the graph. In this way, a curve showing the approximate variation of the cross section of a fish-bone cylinder can be drawn for values of ka' between 0·07 and 1·5 (Fig. 5.42)[14]. The values of ka' for the first three minima are calculated after Faran[16]. Also shown for comparison in Fig. 5.42 is the cross section of a rigid cylinder (taken from Fig. 5.13). σ_G is the normal acoustic cross-section for either of these cases in the geometrical region.

5.24 Universal graphs of cross sections of a fish

A universal graph for targets of the same shape and structure can be obtained by dividing the acoustic cross-section by the square of its length (σ/L^2) and plotting the values so obtained against target length divided by wavelength (L/λ) or frequency multiplied by length (fL). In this way, readings taken by different observers at various frequencies fall on the same curve and may be compared directly. Such Universal graphs for all readings published for fish up to 1965 are seen in Fig. 5.43[35]. Graph (a) is for dorsal aspect and (b) refers to side aspect.

In general, the results of Jones, Pearce and Sothcott taken in fresh water using artificial swimbladders of expanded plastic (to ensure that these structures were of correct size) seem to be somewhat higher than given by the author's work using sticklebacks and minnows without artifice.

Table 5.4 *Acoustic cross-sections of the front surfaces of various parts of a fish (in the geometrical region)*[4]

L = overall length of fish, cm

Part	Approximate shape	Formula	Dimensions	σ side cm²	σ dorsal cm²
Swimbladder	Cylinder	$\frac{2\pi a' l^2 \mathscr{R}^2}{\lambda}$	$a' = 0\cdot0245L$ $l = 0\cdot24L$ $\mathscr{R} = -100\%$	$8\cdot9 \times 10^{-3} L^3/\lambda$	$8\cdot9 \times 10^{-3} L^3/\lambda$
Backbone (core)	Cylinder	$\frac{2\pi a' l^2 \mathscr{R}^2}{\lambda}$	$a' = 0\cdot006L$ $l = 0\cdot65L$ $\mathscr{R} = 24$ to 26%	$1\cdot0 \times 10^{-3} L^3/\lambda$ (mean)	$1\cdot0 \times 10^{-3} L^3/\lambda$
Body	Ellipsoid	Side $\frac{\pi H^2 L_1^2 \mathscr{R}^2}{4B^2}$ Dorsal $\frac{\pi B^2 L_1^2 \mathscr{R}^2}{4H^2}$	$L_1 = 0\cdot93L$ $H = 0\cdot195L$ $B = 0\cdot112L$ $\mathscr{R} = 4\cdot6\%$ (in fresh water) $\mathscr{R} = 1\cdot9\%$ (in sea water)	$4\cdot4 \times 10^{-3} L^2$ (in fresh water) $7\cdot5 \times 10^{-4} L^2$ (in sea water)	$4\cdot7 \times 10^{-4} L^2$ (in fresh water) $8 \times 10^{-5} L^2$ (in sea water)
Alternative body (for long fish at high frequencies)	Plane area	Side aspect only $\frac{4\pi A^2 \mathscr{R}^2}{\lambda^2}$	$A \simeq L_1 H/2$	$2\cdot2 \times 10^{-4} L^4/\lambda^2$ (in fresh water) $3\cdot7 \times 10^{-5} L^4/\lambda^2$ (in sea water)	— —

Cushing's readings using artificial swimbladders are also higher although taken in sea water in which the echo from the body tissue would be reduced, since the reflectivity of the latter $\mathscr{R}$ falls from 4·6 per cent in fresh water to 1·9 per cent in sea water, also the volume of the swimbladder in sea-water species (about 5 per cent of total volume of fish) is lower than in fresh-water types (7 per cent).

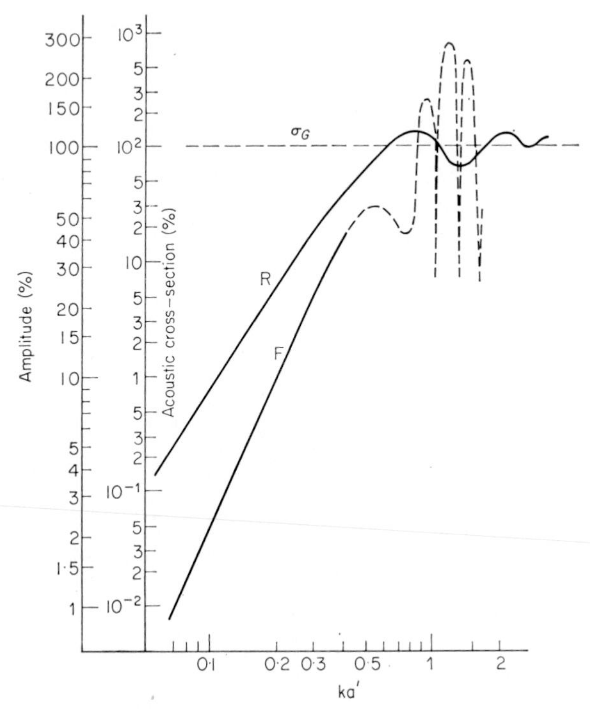

Fig. 5.42 Acoustic cross-sections of rigid (R) and fish bone (F) cylinders in broadside aspect expressed as a percentage of that calculated for the front surface only, in the geometrical region (σ_G), in each case[14]

Clearly from Fig. 5.43(a) and (b) between $L/\lambda = 4$ and $L/\lambda = 18$ there is a region of wide fluctuation in the value of σ (due to interference), with successive maxima and minima. Although the precise form of the graph is difficult to ascertain, some of the minima are more readily picked out (for example in dorsal aspect (a), at $L/\lambda = 12$ and $L/\lambda = 17$).

These experimental results are compared with values in the geometrical region for ellipsoids (lines G, H, L and M), cylinders (lines A and E) and plane targets (lines S and T) calculated from Table 5.4.

To Fig. 5.43 has been added the values in the Rayleigh scattering region for the core of the backbone (F) and a rigid swimbladder (C) (from Fig. 5.42). C represents the lowest-possible amount of energy which could be returned by the swimbladder. The latter is a pressure-release surface but vibrations are damped by the surrounding tissue. In the ideal case of an air bubble surrounded by sea water, a resonance would occur. The properties of a spherical bubble may be examined to give a guide to the performance of cylindrical pressure-release surfaces (which do not appear to have been investigated). It is reasonable to take a bubble of the same volume since the echo-amplitude is proportional to the volume of the target in the Rayleigh scattering region. In the case of fresh-water species, the volume of the swimbladder is 7 per cent of that of the fish, so a sphere of the same volume has a radius $R=5{\cdot}2\times10^{-2}L$. The resonant frequency F_R of this bubble at a pressure of one atmosphere is given by $2\pi R/\lambda=1{\cdot}36\times10^{-2}$ (ref. 5) or $F_R=6\times10^3/L$ in fresh water. (For example, the bladder of a fish of length

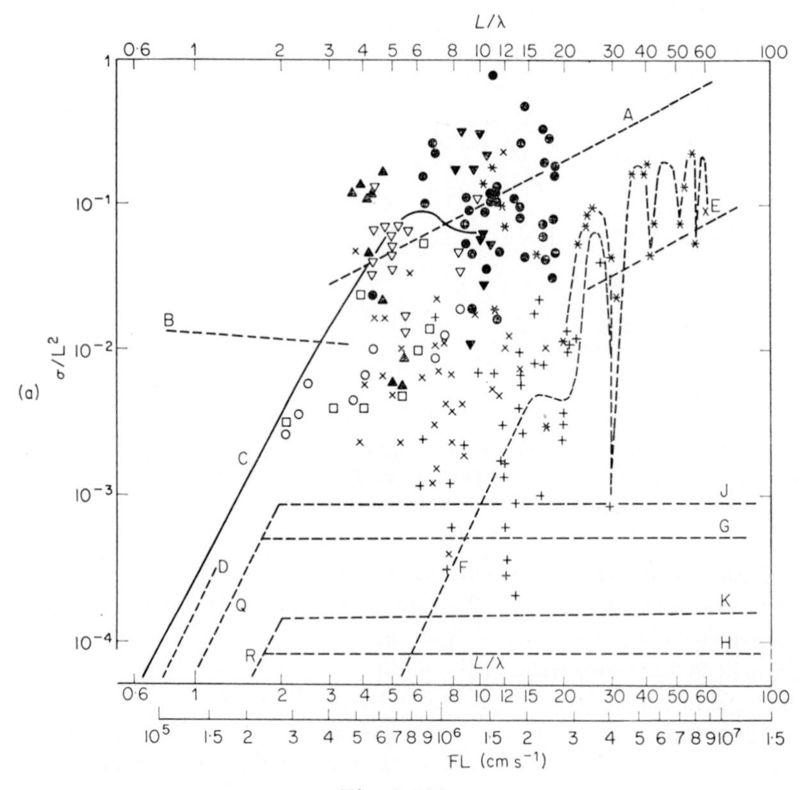

Fig. 5.43(a)

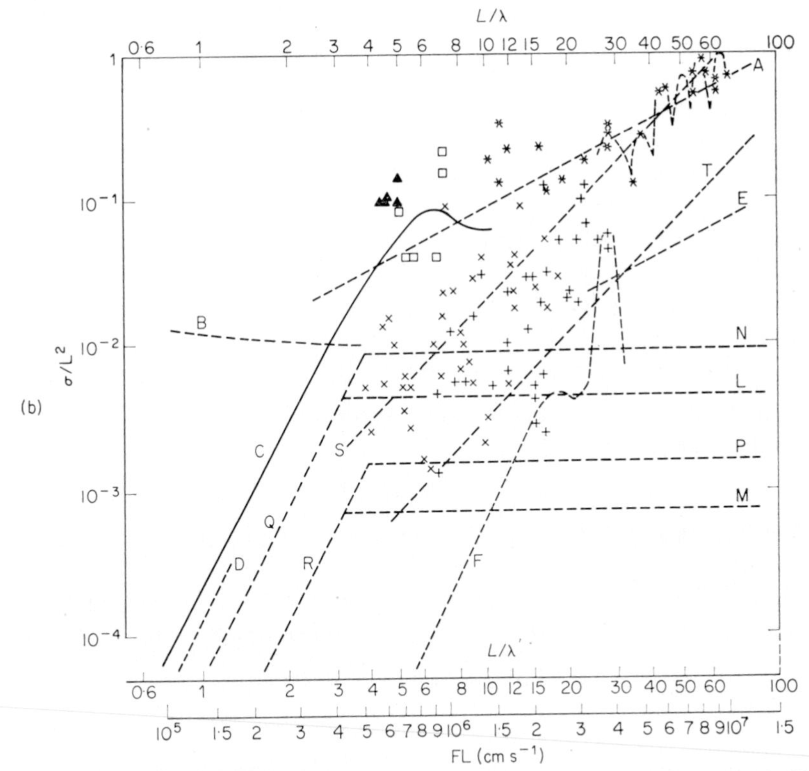

Fig. 5.43 Back-scattering cross-sections σ of fish of various lengths L[35]. Results from several observers are plotted on the universal diagram: (a), dorsal aspect; (b), maximum in side aspect.

□ Hashimoto (28, 300 kHz); △ ▼ Sothcott (15, 30 kHz respectively); ▲ Jones and Pearce (30 kHz); ○ Hashimoto and Maniwa (28, 50, 100 kHz); ● Cushing (30 kHz); × + ✳ Haslett (360, 625 kHz, 1·48 MHz respectively).

The results are compared with the following graphs representing calculated or observed values of σ:

Geometrical region: A, swimbladder; E, core of backbone (front surface only); G, equivalent ellipsoidal body of flesh in fresh water (dorsal aspect, front surface only); H, as for G but in sea water; J, body of flesh in fresh water (dorsal aspect, front surface only, observed); K, as for J but in sea water; L, as for G but in side aspect; M, as for H but in side aspect; N, as for J but in side aspect; P, as for K but in side aspect; S, apparently plane body in side aspect in fresh water; T, as for S but in sea water.

Rayleigh scattering region: B, spherical bubble of same volume as the swimbladder; C, a rigid swimbladder (with its length in geometrical region); D, a rigid swimbladder (completely in Rayleigh scattering region); F, core of backbone complete (observed, with its length in geometrical region); Q, equivalent ellipsoidal body in fresh water; R, as for Q but in sea water

22 cm near the water surface would resonate at about 270 Hz.) This frequency would increase in proportion to the square root of the pressure if neutral buoyancy is maintained by the fish (i.e., constant volume). This resonance corresponds to a value of $L/\lambda \simeq 4 \times 10^{-2}$ in Fig. 5.43 (i.e., off the left-hand side). In the area between this resonance and the geometrical region, σ stays above the geometrical value (viz. $\pi R^2 \simeq 8 \times 10^{-3} L^2$) as indicated by the curve B[32]. The precise position of the curve for the swimbladder of a fish will lie between B and C depending on the 'Q' of the resonance.

For a rigid body having all three dimensions in the Rayleigh scattering region[5]

$$\sigma = 16\pi^3 V^2/\lambda^4 \tag{5.7}$$

where V is the volume of the body. Substituting for the volume of the swimbladder ($V = 5{\cdot}8 \times 10^{-4} L^3$)

$$\sigma = 1{\cdot}7 \times 10^{-4} L^6/\lambda^4$$

which is plotted as line D in Fig. 5.43 and would apply to very small values of L/λ.

Eqn. 5.7 becomes $\sigma = 16\pi^3 V^2 \mathscr{R}^2/\lambda^4$ when the body has an amplitude reflectivity $\mathscr{R}$. Substituting the volume of the fish ($V = 8{\cdot}3 \times 10^{-3} L^3$)

$$\sigma = 7{\cdot}3 \times 10^{-5} L^6/\lambda^4 \text{ in fresh water}$$

$$\sigma = 1{\cdot}2 \times 10^{-5} L^6/\lambda^4 \text{ in sea water}$$

which are plotted as the lines Q and R respectively in Fig. 5.43. In each case, the change from the Rayleigh scattering region to the geometrical region will, of course, be accompanied by an oscillation in the value of the acoustic cross-section (e.g., from line Q to line G).

The lines J, K, N and P are based on the observed measurements on models of the fish body in the geometrical region, in fresh water and sea water respectively, based on an interpolation for fish flesh in Fig. 5.41.

On the universal diagram, the values of acoustic back-scattering cross-section obtained at full scale with fish having artificial swimbladders are rather higher than those in the scale model using very small fish without artifice. The scale-model method shows much more detail concerning the fine structure, including deep minima, between fish lengths of 4 λ and 18 λ.

Acknowledgements

The author is grateful to the following for permission to reproduce figures, tables and passages in the text describing these figures, as listed below.

The Institute of Physics and the Physical Society for Figs. 5.4, 5.10, 5.11–20 inclusive, 5.27, 5.28, 5.30, 5.31, 5.35–43 and Table 5.3.

The Iliffe Press for Figs. 5.2, 5.6, 5.7 and Tables 5.1, 5.2 and 5.4.

The British Institution of Electronic and Radio Engineers for Figs. 5.1, 5.3, 5.5 and 5.26.

The Editor, *Acustica* and the Ministry of Defence (Navy Dept.) for Figs. 5.22 and 5.23.

The Conseil International pour l'Exploration de la Mer for Figs. 5.32–34.

The Editors, *The Journal of Experimental Biology* for Fig. 5.29.

The Acoustical Society of America for Fig. 5.21.

The McGraw-Hill Book Co. for Fig. 5.8.

The Pergamon Press for Fig. 5.9.

The Food and Agricultural Organization of the United Nations, Rome, for Fig. 5.24.

Figs. 5.6, 5.7, 5.11 and 5.22–26 inclusive, appear by courtesy of Kelvin Hughes, a Division of Smiths' Industries Ltd., Hainault, Essex.

References

1 Haslett, R. W. G., 'The Quantitative Evaluation of Echo-sounder Signals from Fish', *J. Brit. I.R.E.* **22**, no. 2, 33–42 (1961).

2 Haslett, R. W. G., and D. Honnor, 'Some Recent Developments in Sideways-looking Sonars', Paper No. 5, *Proc. I.E.R.E.*, Conference on Electronic Engineering in Oceanography, Southampton, September 1966.

3 Haslett, R. W. G., G. Pearce, A. Welsh and K. Hussey, 'The Underwater Acoustic Camera', *Acustica*, **17**, no. 4, 187–203 (1966).

4 Haslett, R. W. G., 'Physics Applied to Echo Sounding for Fish', *Ultrasonics*, Jan.–Mar., 11–22 (1964).

5 N.D.R.C. Division 6, Washington, 'Physics of Sound in the Sea', *Summary Technical Report*, Vol. 8, pp. 16, 105, 347, 361 and 463 (1946).

6 Schulkin, M., and H. W. Marsh, 'Absorption of Sound in Sea Water', *J. Brit. I.R.E.* **25**, no. 6, 493 (1963).

7 Sünd, O. 'The Fat and Small Herring on the Coast of Norway in 1940', Rapp. Cons. Explor. Mer., *Ann. Biol.* **1**, 62 (1939–41).

8 Haslett, R. W. G., 'The Back-scattering of Acoustic Waves in Water by an Obstacle—I: Design of a Scale Model and Investigation of its Validity', *Proc. Phys. Soc.* **79**, 542–558 (1962).

9 Mentzer, J. R., *Scattering and Diffraction of Radio Waves*, p. 31, Fig. 11 p. 64, pp. 76, 104, 131, Pergamon, Oxford, 1955.

10 Kerr, D. E., 'Propagation of Short Radio Waves', *Radn. Lab. Ser. No. 13*, 1st edition, Fig. 6.1 p. 453, pp. 457, 461, 464, McGraw-Hill, New York, 1951.

11 Haslett, R. W. G., 'Back-scattering of Acoustic Waves in Water by an Obstacle—II: Determination of the Reflectivities of Solids Using Small Specimens', *Proc. Phys. Soc.* **79**, 559–571 (1962).

12 Harden-Jones, F. R., 'Results of Echo-sounding Experiments on Single Fish—R. V. Ernest Holt—Cruise VI/1955' (Fisheries Laboratory, Lowestoft, personal communication).
13 Hopkin, P. R., 'Cathode-ray Tube Displays for Fish Detection on Trawlers', *J. Brit. I.R.E.* **25**, no. 1, 77 (1963).
14 Haslett, R. W. G., 'The Acoustic Back-scattering Cross-sections of Short Cylinders', *Brit. J. Appl. Phys.* **15**, 1085–94 (1964).
15 Sothcott, J. E. L., Kelvin Hughes, Hainault, Essex, personal communication.
16 Faran, J. J., 'Sound Scattering of Solid Cylinders and Spheres', *J. Acoust. Soc. Am.* **23**, no. 4, 405 (1951).
17 Haslett, R. W. G., 'Acoustic Backscattering from an Air-filled Cylindrical Hole Embedded in a Sound-translucent Cylinder', *Brit. J. Appl. Phys.* **17**, 549–561 (1966).
18 Aldridge, E. E., 'An Investigation into the Mode of Working of the Ultrasonic Micrometer Used in Thickness Measurements of Thin Wall Steel Tubing', *A.E.R.E. Report* M. 1510, 1965.
19 Rayleigh, Lord, *Theory of Sound*, Vol. 2, p. 88, Macmillan, London, 1929.
20 Fay, R. D., and O. V. Fortier, 'Transmission of Sound Through Steel Plates Immersed in Water', *J. Acoust. Soc. Am.* **23**, no. 3, pp. 339–346 (1951).
21 Haslett, R. W. G., 'An Ultrasonic to Electronic Image Converter Tube for Operation at 1·2 Mc/s', *Radio Electron. Eng.* **31**, no. 3, pp. 161–170 (1966).
22 Ellis, G. H., P. R. Hopkin and R. W. G. Haslett, *F.A.O. Second World Fishing Gear Congress*, London, 1963, Paper No. 5, Fig. 3.
23 Ellis, G. H., P. R. Hopkin, and R. W. G. Haslett, *Fishing Gear of the World* 2 *—A Comprehensive Echo-sounder for Distant Water Trawlers*, p. 365, Fishing News (Books) Ltd., London, 1964.
24 Haslett, R. W. G., 'A High-speed Echo-sounder Recorder having Seabed Lock', *J. Brit. I.R.E.* **24**, pp. 441–452 (1962).
25 Haslett, R. W. G., 'Determination of the Acoustic Back-scattering Patterns and Cross Sections of Fish', *Brit. J. Appl. Phys.* **13**, pp. 349–357 (1962).
26 Jones, F. R. H., and G. Pearce, 'Echo Sounding Experiments with Perch to Determine the Proportion of the Echo Returned by the Swim Bladder', *J. Expl Biol.* **35**, no. 2, 437 (1958).
27 Hashimoto, T., *Report of Fishing Boat Lab. No. 1*, Fishing Agency, Ministry of Agriculture and Forestry, Tokyo, Japan, 1953.
28 Hashimoto, T., and Y. Maniwa, 'Study of the Reflection Loss of Ultrasonic Wave on Fish Body by Millimetre Wave', *Technical Report No. 8*, p. 113, Fishing Boat Lab., Ministry of Agriculture and Forestry, Tokyo, Japan, 1956.
29 Shishkova, E. V., 'Study of Acoustical Characteristics of Fish', *F.A.O. Second World Fishing Gear Congress*, London (May, 1963). Paper No. 74.
30 Cushing, D. H., and I. D. Richardson, 'A Triple Frequency Echo Sounder', *Ministry of Agriculture and Fisheries Fishery Investigations*, Ser. 2 **20**, 1 (1955).
31 McCartney, B. S., National Institute of Oceanography, Wormley, Surrey, personal communication.
32 Devin, C., 'Survey of Thermal Radiation and Viscous Damping of Pulsating Air Bubbles in Water', *J. Acoust. Soc. Am.*, **31** 1654–1667 (1959).

33 Haslett, R. W. G., 'Measurements of the Dimensions of Fish to Facilitate Calculations of Echo-strength in Acoustic Fish Detection', *J. du Cons. Intern. l'explor. mer* **27**, no. 3, 261–269 (1962).

34 Haslett, R. W. G., 'Determination of the Acoustic Scatter Patterns and Cross Sections of Fish Models and Ellipsoids', *Brit. J. Appl. Phys.* **13**, 611–620 (1962).

35 Haslett, R. W. G., 'Acoustic Backscattering Cross Sections of Fish at Three Frequencies and their Representation on a Universal Graph', *Brit. J. Appl. Phys.* **16**, 1143–1150 (1965).

6

The Non-linear Interaction of Acoustic Waves

V. G. Welsby
Electrical Engineering Department, University of Birmingham

6.1 Theory

6.1.1 Introduction

An exact mathematical analysis of finite-amplitude acoustic waves (so called to distinguish them from the waves of infinitesimal amplitude which are assumed in the derivation of the classical acoustic wave equation) is difficult since it must necessarily be concerned with the solution of non-linear differential equations. The theory of finite-amplitude waves has been well covered in the literature and various accepted methods are used to obtain approximate solutions which are usually adequate for the particular purposes for which they are intended. In this lecture, I propose to trace the main lines of argument on which the theory is based, placing the emphasis on the physical meaning of the various steps. The analysis will be presented from the point of view of the design engineer rather than the physicist but experience has shown that, in this subject, an adequate grasp of fundamental theory is essential if the potential practical applications are to be properly understood.

6.1.2 *An intuitive approach and its limitations*

The inherent non-linearity of acoustic waves can be demonstrated by the following line of thought, although, at the same time, this section can serve as a warning that apparent analogies may sometimes be misleading.

Compare, to start with, the differential equations for the pressure and particle velocity in a uniform plane acoustic wave of infinitesimal amplitude with those for the voltage and current in a uniform electrical transmission line.

$$\frac{\partial p}{\partial x} = -\rho \frac{\partial u}{\partial t} \tag{6.1}$$

$$\frac{\partial u}{\partial x} = -\frac{1}{A}\frac{\partial p}{\partial t} \tag{6.2}$$

$$\frac{\partial V}{\partial x} = -L\frac{\partial I}{\partial t} \tag{6.3}$$

$$\frac{\partial I}{\partial x} = -C\frac{\partial V}{\partial t} \tag{6.4}$$

It seems reasonable to think of pressure, p, and velocity, u, as the analogues of voltage, V, and current, I, respectively, and to note the existence of very similar expressions for the wave impedance, Z_0, and velocity of propagation, C_0, in the two cases. L and C respectively represent the inductance and capacitance per unit length of the electrical transmission line while ρ is the density and A the appropriate elastic modulus of the medium.

$$C_0 = \sqrt{\frac{A}{\rho}} \tag{6.5}$$

$$Z_0 = \sqrt{\rho A} = \rho C_0 \tag{6.6}$$

$$C_0 = \frac{1}{\sqrt{LC}} \tag{6.7}$$

$$Z_0 = \sqrt{\frac{L}{C}} \tag{6.8}$$

Density and compressibility then appear as the analogues of inductance and capacitance, respectively. It can be argued that, since the velocity of propagation and the impedance of acoustic waves each depend both on the density and compressibility of the medium, then they must be affected by anything which changes the values of these elastic properties of the medium. But the passage of an acoustic wave evidently causes the density to fluctuate and it is easy to see that it must also affect the compressibility, $1/A$. Thus it is not surprising to find that non-linear effects became apparent when the wave amplitude is sufficiently large for these fluctuations to become significant.

It can also be noted that electrical transmission lines can be constructed in which the values of L and C are functions of I and V.

So far this argument is perfectly valid; the danger comes if what appears to be the next logical step is taken. That is, if it is automatically assumed that an electrical analogue for finite-amplitude acoustic waves can be constructed simply by finding the dependence of $1/A$ and ρ on pressure and velocity and then translating this into a corresponding dependence of C and L on voltage and current. This is not to say that it would be impossible to find an equivalent electrical transmission line to represent the propagation of uniform plane acoustic waves of finite amplitude but that its form might not be quite as expected. The point is that, in spite of the similarity of the linear equations for infinitesimal waves in the two media, the ways in which these equations depart from linearity as the wave amplitude is increased are different. To explain why this is so it will be necessary to go through a more exact analysis of the type to be described in the next section. A clue is provided however by the thought that, when considering the velocity of the particles, we have to be careful to state the frame of reference with respect to which these velocities are to be measured. This may lead to a kind of ambiguity which does not arise in any ordinary electrical transmission line.

The idea of comparing acoustic waves with electromagnetic waves along a transmission line will be taken up again briefly in Section 6.1.5.

6.1.3 Basic theory

The method adopted is the so-called 'quasi-linear' or 'source-density' one, which is used generally as a convenient way of obtaining approximate solutions to field problems defined by differential equations which are non-linear, but only slightly so. The method was introduced to acoustical theory by Westervelt[1] who used it obtain his well-known 'source-density' equation relating to the interaction of two collimated plane-wave beams travelling in the same direction.

The source-density method is based on the following argument. To start with, it is assumed that the interaction between the primary waves is so slight that no appreciable error is introduced by treating each field as if the other were not there. The second-order interaction products, in each element of volume, are then treated as if they were due to sources existing within that element. Finally, the resulting field is computed by pretending that the virtual sources behave like real sources and that the wavelets radiated from them combine to form the resultant, without any further interaction between themselves or with the primary waves. There are two situations where these assumptions may not be justified; one occurs when the wave amplitudes are so great that the threshold of cavitation is approached. The other occurs when the distance, over which the cumulative effect of the interaction of high-intensity waves is integrated, approaches the 'discontinuity distance' where the actual change in the shape of the wave becomes significant. At the discontinuity distance, for example, an initially sinusoidal wave begins to assume a 'sawtooth' shape. One way of looking at this effect is to attribute it to interaction between each wave and itself. Unless otherwise stated, these limiting conditions will be excluded for the present purpose and it will be assumed that the 'quasi-linear' assumptions are valid.

We start with the two equations defining acoustic waves in a non-viscous fluid. As mentioned above, it is most important to be clear about the way in which the velocity of the particles is defined. It must be stated therefore that these equations are defined for an 'Eulerian' coordinate system, i.e., one which is stationary with respect to the undisturbed fluid in the absence of any acoustic waves

$$\nabla p = -\rho \frac{\mathrm{d}}{\mathrm{d}t} \boldsymbol{u} \tag{6.9}$$

$$\nabla . (\boldsymbol{u}\rho) = -\frac{\partial \rho}{\partial t} \tag{6.10}$$

The first is the equation of motion which relates the pressure gradient to the product of the mass of fluid per unit volume, multiplied by the vector acceleration of the particles. The second is the continuity equation which relates the divergence of the vector rate of mass-flow (i.e., the net rate of mass-flow out of a fixed unit volume) to the rate of change of density of the fluid within that volume element. In addition there is the equation of state which relates the differential pressure p in any volume element to the change of density, from its initial value ρ_0 with the acoustic field absent, of the fluid within that volume. This can be expressed as a power series of the form

$$p = A\left(\frac{\rho - \rho_0}{\rho_0}\right) + \frac{B}{2}\left(\frac{\rho - \rho_0}{\rho_0}\right)^2 + \dots \tag{6.11}$$

where the first two coefficients A and B have the following values

$$A=\rho_0\left(\frac{\partial p}{\partial \rho}\right)_0 \tag{6.12}$$

$$B=\rho\rho_0\left(\frac{\partial^2 p}{\partial \rho^2}\right)_0\simeq\rho_0^2\left(\frac{\partial^2 p}{\partial \rho^2}\right) \tag{6.13}$$

By rearranging Eqn. 6.11, differentiating it and retaining only first- and second-order terms, the following relationships are obtained

$$\rho=\rho_0\left(1+\frac{p}{A}\right) \tag{6.14}$$

$$\frac{\partial \rho}{\partial p}=\frac{\rho_0}{A}\left(1-\frac{B}{A^2}p\right) \tag{6.15}$$

By suitable manipulation of Eqns. 6.12–15, the following second-order approximations to Eqns. 6.9 and 6.10 can be derived (see Appendix I, Eqns. 6A.3 and 6A.4)

$$\nabla p=-\rho_0\frac{\partial}{\partial t}\boldsymbol{u}+\tfrac{1}{2}\nabla\left[\frac{p^2}{A}-\rho_0 u^2\right] \tag{6.16}$$

$$\nabla.\boldsymbol{u}=-\frac{1}{A}\frac{\partial p}{\partial t}+\frac{1}{2A}\frac{\partial}{\partial t}\left[\left(1+\frac{B}{A}\right)\frac{p^2}{A}+\rho_0 u^2\right] \tag{6.17}$$

It is also shown in Appendix I that these can be combined to give a non-linear wave equation of the form

$$\nabla^2 p=\frac{1}{C_0^2}\frac{\partial^2}{\partial t^2}\left[p-\frac{1}{A}\left(\frac{B}{2A}p^2+Z_0 u^2\right)\right] \tag{6.18}$$

Note that, if the amplitudes of p and u are sufficiently small, Eqns. 6.16–18 reduce to the classical equations for infinitesimal waves.

$$\nabla p=-\rho_0\frac{\partial}{\partial t}\boldsymbol{u} \tag{6.19}$$

$$\nabla.\boldsymbol{u}=-\frac{1}{A}\frac{\partial p}{\partial t} \tag{6.20}$$

$$\nabla^2 p=\frac{1}{C_0^2}\frac{\partial^2}{\partial t^2}p \tag{6.21}$$

6.1.4 'Source-density' theory

It is of course the term involving p^2 and u^2 in Eqn. 6.18 which leads to non-linear effects for waves of finite amplitude. As has already been pointed out, exact solutions of differential equations such as Eqn. 6.18 are difficult to find so that various artifices have to be used in order to obtain simple approximate solutions which are sufficiently accurate for practical purposes under certain specified conditions. One such method of approach is the so-called 'quasi-linear' or 'source-density' one which is based on the following argument. To start with, it is assumed that the interaction between the primary waves is so slight that no appreciable error is introduced by computing each wave as though the other were not there. The second-order products, in each element of volume, are then treated as if they were due to appropriate virtual sources existing within that volume. Finally, the resulting field is computed by pretending that the virtual sources behave like real sources and that the wavelets radiated from them combine to form the resultant field without any further interaction with each other or with the primary waves.

Eqn. 6.18 can be expressed in the form

$$\nabla^2 p = \frac{1}{C_0^2}\frac{\partial^2 p}{\partial t^2} + \nabla^2 p' \tag{6.22}$$

where

$$\nabla^2 p' = -\frac{1}{C_0^2}\frac{\partial^2}{\partial t^2}\left[\frac{B}{2A^2}p^2 + \frac{(Z_0 u)^2}{A}\right] \tag{6.23}$$

It remains to be shown that this additional term in the wave equation is to be interpreted as a source density and to find the source strength. This can be done by noting that a source of strength q per unit volume is, by definition, equal to a divergence of the particle velocity vector field. That is

$$q = \nabla . \boldsymbol{u}$$

Consider Eqn. 6.9

$$\nabla p = -\rho \frac{\mathrm{d}}{\mathrm{d}t}\boldsymbol{u}$$

$$\simeq -\rho_0 \frac{\partial}{\partial t}\boldsymbol{u}$$

Take the divergence of both sides

$$\nabla . \nabla p = \nabla^2 p \simeq -\rho_0 \frac{\partial}{\partial t}\nabla . \boldsymbol{u}$$

whence

$$\nabla_0 \boldsymbol{u} \simeq -\frac{1}{\rho_0}\int \nabla^2 p \mathrm{d}t = q$$

It follows that the additional component of $\nabla^2 p$, denoted by $\nabla^2 p'$ in Eqn. 6.22, is equivalent to a source strength density

$$q=-\frac{1}{\rho_0}\int \nabla^2 p' \mathrm{d}t \tag{6.24}$$

$$=\frac{1}{A^2}\frac{\partial}{\partial t}\left[\frac{B}{2A}p^2+(Z_0 u)^2\right] \tag{6.25}$$

$$=\frac{1}{\rho_0^2 C_0^4}\frac{\partial}{\partial t}\left[\frac{B}{2A}p^2+(Z_0 u)^2\right] \tag{6.26}$$

since

$$A=\rho_0 C_0^2 \tag{6.27}$$

and

$$Z_0^2=\frac{\rho_0}{A} \tag{6.28}$$

Eqn. 6.26 gives the source strength density for primary fields of any configuration. Note that this is a function of both $(\partial/\partial t)p^2$ and $(\partial/\partial t)u^2$. Simpler expressions involving only $(\partial/\partial t)p^2$ are applicable in special cases where there happens to be a constant ratio between p and $Z_0 u$.

6.1.5 *Equivalent electrical transmission lines*

Let us now return to the idea hinted at in Section 6.1.2 and consider the possibility of finding an electrical transmission line analogue to represent uniform plane acoustic waves. The 'one-dimensional' form of Eqns. 6.16 and 6.17 are

$$\frac{\partial p}{\partial x}=-\rho_0\frac{\partial u}{\partial t}+\frac{1}{2}\frac{\partial}{\partial x}\left[\frac{p^2}{A}-\rho_0 u^2\right] \tag{6.29}$$

$$\frac{\partial u}{\partial x}=-\frac{1}{A}\frac{\partial p}{\partial t}+\frac{1}{2A}\frac{\partial}{\partial t}\left[\left(1+\frac{B}{A}\right)\frac{p^2}{A}+\rho_0 u^2\right] \tag{6.30}$$

But $A=\rho_0 C_0^2$, so that $\rho_0 A_0=Z_0^2$ and, for plane waves

$$u^2=\frac{p^2}{Z_0^2}=\frac{p^2}{\rho_0 A}$$

Thus the equations become

$$\frac{\partial p}{\partial x}=-\rho_0\frac{\partial u}{\partial t} \tag{6.31}$$

$$\frac{\partial u}{\partial x}=-\frac{1}{A}\frac{\partial}{\partial t}(p-\eta p^2) \tag{6.32}$$

where

$$\eta=\frac{1}{A}\left(1+\frac{B}{2A}\right) \tag{6.33}$$

For comparison, suppose that the capacitance C in Eqn. 6.4 is of the form $C=C_0(1-2kV)$ where k is a constant.
Then

$$\frac{\partial I}{\partial x}=-C_0\frac{\partial V}{\partial t}+2kC_0V\frac{\partial V}{\partial t} \tag{6.34}$$

$$=-C_0\frac{\partial}{\partial t}(V-kV^2) \tag{6.35}$$

Comparison of Eqns. 6.31, 6.32 and 6.3, 6.4 implies that the equivalent transmission line should have a constant inductance per unit length, all the non-linearity being represented by the dependence of the capacitance per unit length on the voltage across the line.

This is not quite what might have been expected because the density is known to fluctuate and this would suggest a changing inductance in the electrical analogue. The physical explanation of this apparent paradox is bound up with the fact that acoustic waves, unlike electromagnetic ones, involve positional displacement of the medium through which they travel. What Eqn 6.31 tells us is that, for uniform plane waves

$$\rho\frac{\mathrm{d}u}{\mathrm{d}t}=\rho_0\frac{\partial u}{\partial t}$$

In other words, the variation of density of the portion of the medium within a given element of volume in a fixed frame of reference is exactly cancelled out by the motion of the medium so that the density, as seen by a fixed observer, remains constant.

Note that these remarks apply only for a single wave or for a group of waves travelling in the same direction. The equivalent transmission line already discussed will not represent correctly the interaction of a pair of waves travelling in opposite directions. This rules out the possibility of using an electrical analogue to demonstrate acoustic wave interaction near a reflecting boundary, that is where the primary field contains incident and reflected waves which are travelling in opposite directions.

As a matter of interest, it is possible to find an equivalent line for waves in opposite directions; the point is that it cannot be the *same* line as for waves in the same direction. It is shown in Appendix II that, for waves in opposite directions and as far as the interaction products only are concerned, ρ_0u^2 is equivalent to $-p^2/A$, so that Eqns. 6.29 and 6.30 then become

$$\frac{\partial p}{\partial x}=-\rho_0\frac{\partial u}{\partial t}+\frac{2p}{A}\frac{\partial p}{\partial x} \tag{6.36}$$

$$\simeq -\rho_0\left(1+2\frac{p}{A}\right)\frac{\partial u}{\partial t} \quad (6.37)$$

$$\frac{\partial u}{\partial x} = -\frac{1}{A}\frac{\partial p}{\partial t} + \frac{B}{A^3}p\frac{\partial p}{\partial t} \quad (6.38)$$

$$\simeq -\frac{1}{A}\left(1-\frac{B}{A^2}p\right)\frac{\partial p}{\partial t} \quad (6.39)$$

Comparison with the equations for the transmission line now show that not only must the inductance per unit length vary as well as the capacitance but the inductance must depend on the voltage across the line rather than the current. This point is probably only of academic interest; the only reason for including it is to underline the misunderstandings which can occur if the apparent analogy between acoustic and electromagnetic waves is carried too far.

6.1.6 *Summary*

As far as practical applications of the theory are concerned, probably the most important result is that the non-linear wave equation is generally of the form

$$\nabla^2 p = \frac{1}{C_0^2}\frac{\partial^2}{\partial t^2}\left[p - \frac{1}{A}\left(\frac{B}{2A}p^2 + (Z_0 u)^2\right)\right]$$

leading to a general source-density function

$$q = \frac{1}{A^2}\frac{\partial}{\partial t}\left[\frac{B}{2A}p^2 + (Z_0 u)^2\right]$$

But if, *and only if*, we are concerned with plane waves travelling in the same direction, the equations can be simplified to

$$\frac{\partial^2 p}{\partial x^2} = \frac{1}{C_0^2}\frac{\partial^2}{\partial t^2}[p - \eta p^2]$$

where

$$\eta = \frac{1}{A}\left(1+\frac{B}{2A}\right)$$

$$q = \frac{1}{A^2}\left(1+\frac{B}{2A}\right)\frac{\partial}{\partial t}p^2$$

These results can be used as the basis for the study of possible devices and systems utilizing the non-linear interaction of acoustic waves.

6.2 Applications

6.2.1 *Endfire array of virtual sources*

Westervelt[1] showed that a pair of superimposed plane waves, in the form of an idealized collimated beam, will produce source distributions, at the sum and difference frequencies, which behave as virtual arrays capable of radiating power at these frequencies. A simple way of seeing that this is so is to start by considering the effect of a single volume element within a narrow beam, as observed at a distant receiver situated on the beam axis. The total delay, obtained by adding together the delay of the primary waves up to the element and the delay of the scattered waves beyond it, will be independent of the position of the element. Thus the contributions from all the sources will add together at the receiver in phase. The integrated effect of all the sources will be a maximum for an observer on the axis but will fall off quite rapidly as the latter moves away from the axis because the delays will no longer be independent of the element positions and partial cancellation will occur. The arrangement will behave as an endfire array. The source strength at the sum and difference frequencies, since they arise from the cross-product of the two primary waves, will fall with distance at an exponential rate $e^{-(\alpha_1+\alpha_2)}$ where α_1 and α_2 are the respective attenuation coefficients at the primary frequencies. The virtual array will thus have an exponential 'taper'. Since the collimated beam must have a finite width b, the directional pattern for an ideal line array must be multiplied by the diffraction pattern for an aperture of width b, so that the total directional function is of the form

$$D(\theta)=\left[1+\left(\frac{2K}{\alpha}\sin^2\frac{\theta}{2}\right)^2\right]^{-1/2}\frac{\sin X}{X} \tag{6.40}$$

where $\alpha=\alpha_1+\alpha_2$, $K=(k_1-k_2)$ or (k_1+k_2) and $X=bK\sin\theta$. The expression in the brackets [ ] is the directional function for the ideal endfire line array with an exponential taper coefficient α. The significant range of θ will usually be confined to values for which $\sin\theta\simeq\theta$, so that

$$D(\theta)\simeq\left[1+\left(\frac{K\theta^2}{2\alpha}\right)^2\right]^{-1/2}$$

The 'half-power' beamwidth of the directional pattern will then be defined by $\pm\theta_d$ where

$$\theta_d=2\sqrt{\frac{\alpha}{2K}}$$

So far it has been assumed that the primary waves retain the form of a collimated beam over the significant distance, that is until their amplitude has been attenuated to a negligible level. Berktay[2] has however developed this idea of a virtual endfire array further and has obtained solutions for cases where the primary waves spread cylindrically or spherically, in the form of beams having constant angles in one or two planes, instead of being of constant width. His results, for the difference frequency, for a cylindrically-spreading 'fan' beam are indicated in Fig. 6.1. The additional

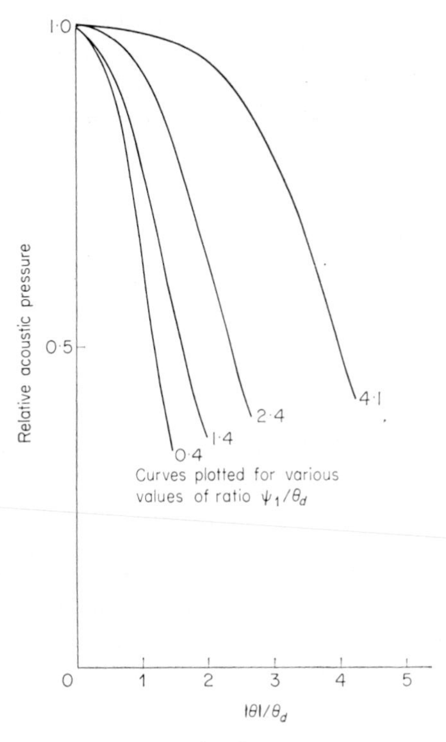

Fig. 6.1

variable here is the half-angle ψ_1 of the primary fan beam. It appears in the curves in the form of a parameter ψ_1/θ_d where θ_d has the same meaning as before. It will be observed that, if ψ_1 is less than θ_d, the result is still practically independent of the aperture but, as soon as ψ_1 exceeds θ_d, the beam-width at the difference frequency becomes progressively larger as ψ_1 is increased.

It is interesting to note the effect of changing the value of θ_d, for example by altering the difference frequency while keeping the total attenuation coefficient practically constant. The resulting change in the horizontal scale of the curves can be made to offset the effect of the change in the ratio ψ_1/θ_d so that the beamwidth will tend to remain constant.

Thus the virtual array technique not only provides a method of forming a narrow transmitted beam with an aperture much smaller than would be required to produce the same beamwidth by direct transmission, but the beamwidth can be kept constant over a much wider frequency band.

A general point which must be mentioned is that since, under given conditions, the effective radiated power is proportional to the square of the total power input to the system, there will be a tendency to operate at the highest practicable power level. This may mean that significant non-linear interaction of each primary wave with *itself* will occur. In other words, the so-called 'discontinuity distance' for each primary wave, beyond which its waveform assumes a 'sawtooth' rather than a sinusoidal shape, may occur within the effective length of the exponential array. The increased attenuation caused by this will cause some widening of the beam because it reduces the active length of the array.

6.2.2 Parametric receiving array

The basic idea of cumulative interaction between waves travelling in the same direction can be applied to reception as well as transmission. In this case, the interaction takes place between the incoming signal waves and a local, high-intensity 'pump' wave. The receiver can then be tuned to pick up the sum or difference wave. As in the transmitting array described in Section 6.2.1, the main object of an arrangement of this kind would be to produce high directivity, using physical transducers smaller than those which would be required to achieve the same result to direct reception. The remarks in the previous section about waveform distortion and increased attenuation of high-intensity waves will apply here to the pump wave. Theoretical analysis is also complicated by the fact that, when dealing with the interaction of two waves of vastly differing power levels, the basic assumption of no secondary interaction between the scattered and primary waves may not be fully justified. The practical usefulness of parametric receivers of this type is still under investigation[3].

6.2.3 Other applications

Interaction will take place between waves even when their directions of transmission are different (see Appendix II). The practical usefulness of this fact is however limited because, from simple geometrical considerations, the region of interaction is likely to be relatively small. Furthermore the phase

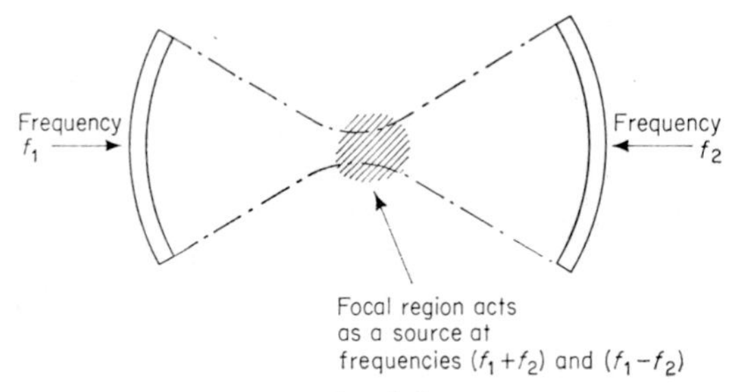

Fig. 6.2

distribution of the source density is such that mutual cancellation of the radiated wavelets tends to occur rather than the desired cumulative summation.

An interesting situation arises however if the primary waves are focused into a common focal region (see Fig. 6.2). Because of the 'square-law' property of the non-linearity, the significant interaction will be mainly confined to the focal region, which then behaves as a virtual source at the sum and difference frequencies.

We thus have a method of producing a virtual source at a point at some distance from any physical transducer. By a suitable choice of frequencies, the pair of focusing transducers can be made to behave as a resonant cavity at the difference frequency, thus enhancing the output at that frequency and giving the whole device some measure of directivity[4].

There is another application for the same arrangement; which is as a 'probe' to detect changes in the acoustic properties of the medium in the focal region. Suppose, for example, that a single focusing transducer is used with a single primary frequency and a separate small transducer is used to pick up the second-harmonic wave radiated from the focal region

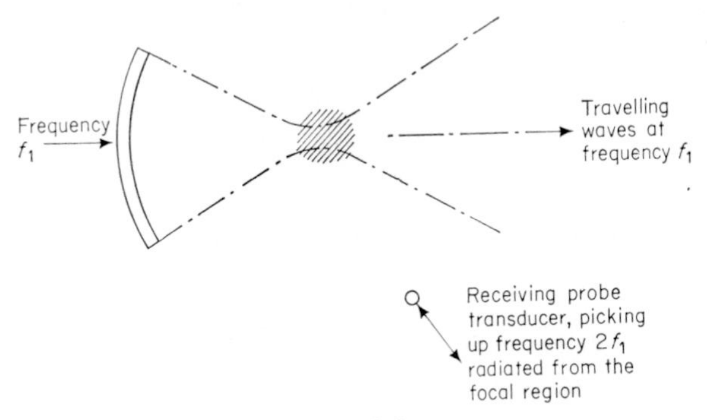

Fig. 6.3

(Fig. 6.3). That is, we make use of the interaction between a single wave and itself. It has been demonstrated[5] that changes in the amplitude and phase of the harmonic component occur just before the cavitation threshold is reached, presumably caused by changes in the acoustic properties of the medium in the focal region where the acoustic intensity is highest. This effect may be of practical use as a means of detecting the approach of cavitation without actually letting breakdown of the liquid occur.

Appendix I

Derivation of second-order acoustic equations

We start with the two equations which must be satisfied in a non-viscous fluid.

$$\nabla p = -\rho \frac{\mathrm{d}}{\mathrm{d}t}\boldsymbol{u} \tag{6.9}$$

$$\nabla . (\boldsymbol{u}\rho) = -\frac{\partial \rho}{\partial t} \tag{6.10†}$$

The RHS of Eqn. 6.9 contains the full derivative of the vector particle velocity $\boldsymbol{u}$ with respect to time. This can be split into two parts, one of which is the partial rate of change of $\boldsymbol{u}$ at a fixed position, i.e., $(\partial/\partial t)\boldsymbol{u}$. The component of the other part in each coordinate direction is the product of the rate of change of $\boldsymbol{u}$ with respect to displacement in that direction, multiplied by the actual rate of change of displacement in that direction as a function of time. Thus

$$\frac{\mathrm{d}}{\mathrm{d}t}\boldsymbol{u} = \frac{\partial}{\partial t}\boldsymbol{u} + \left(u_x \frac{\partial}{\partial x} + u_y \frac{\partial}{\partial y} + u_z \frac{\partial}{\partial z}\right)(u_x\boldsymbol{i} + u_y\boldsymbol{j} + u_z\boldsymbol{k})$$

or, written in vector notation

$$\frac{\mathrm{d}}{\mathrm{d}t}\boldsymbol{u} = \frac{\partial}{\partial t}\boldsymbol{u} + (\boldsymbol{u} . \nabla)\boldsymbol{u}$$

Making use of an identity from vector algebra, we get

$$\frac{\mathrm{d}}{\mathrm{d}t}\boldsymbol{u} = \frac{\partial}{\partial t}\boldsymbol{u} + \frac{1}{2}\nabla u^2 - \boldsymbol{u} \times (\nabla \times \boldsymbol{u})$$

where u is the magnitude of the $\boldsymbol{u}$ vector, i.e.

$$u^2 = u_x^2 + u_y^2 + u_z^2$$

A simplification can be made since, in a non-viscous fluid, the field can be assumed to be irrotational. This means that the quantity $(\nabla \times \boldsymbol{u})$, which

† Equation numbers not containing 'A' refer to the main text of this chapter.

represents the curl of the velocity vector, is zero. Therefore the last term in the expression for $(\mathrm{d}/\mathrm{d}t)\boldsymbol{u}$ is also zero and we have

$$\frac{\mathrm{d}}{\mathrm{d}t}\boldsymbol{u}=\frac{\partial}{\partial t}\boldsymbol{u}+\frac{1}{2}\nabla u^2 \tag{6A.1}$$

Making use of this relationship and also of Eqn. 6.14 we can now put Eqn. 6.9 into the form

$$\nabla p=-\rho_0\left(1+\frac{p}{A}\right)\left(\frac{\partial}{\partial t}\boldsymbol{u}+\frac{1}{2}\nabla u^2\right)$$

Expanding and retaining only first- and second-order terms gives

$$\nabla p=-\rho_0\frac{\partial}{\partial t}\boldsymbol{u}-\frac{\rho_0}{A}p\frac{\partial}{\partial t}\boldsymbol{u}-\frac{1}{2}\rho_0\nabla u^2 \tag{6A.2}$$

Now consider the quantity $p(\partial/\partial t)\boldsymbol{u}$. This is already a second-order term so that only a negligible error will be introduced if, within it, we make use of the linear approximation to Eqn. 6A.2 obtained by retaining only the first-order terms, i.e.

$$\nabla p\simeq-\rho_0\frac{\partial}{\partial t}\boldsymbol{u}$$

or

$$\frac{\partial}{\partial t}\boldsymbol{u}\simeq-\frac{1}{\rho_0}\nabla p$$

Substituting this into the second term of Eqn. 6A.2 gives

$$\begin{aligned}\nabla p&=-\rho_0\frac{\partial}{\partial t}\boldsymbol{u}+\frac{p}{A}\nabla p-\frac{1}{2}\rho_0\nabla u^2\\&=-\rho_0\frac{\partial}{\partial t}\boldsymbol{u}+\frac{1}{2A}\nabla p^2-\frac{1}{2}\rho_0\nabla u^2\\&=-\rho_0\frac{\partial}{\partial t}\boldsymbol{u}+\frac{1}{2}\nabla\left[\frac{p^2}{A}-\rho_0u^2\right]\end{aligned} \tag{6A.3}$$

Eqn. 6A.3 is the required second-order approximation to the original equation of motion, Eqn. 6.9.

Next, consider the continuity equation, Eqn. 6.10

$$\begin{aligned}\nabla.(\rho\boldsymbol{u})&=-\frac{\partial\rho}{\partial t}\\&=-\frac{\partial p}{\partial t}\cdot\frac{\partial\rho}{\partial p}\end{aligned}$$

Substituting from Eqns. 6.14 and 6.15

$$\rho_0 \nabla . \boldsymbol{u} + \frac{\rho_0}{A} \nabla . (p\boldsymbol{u}) = -\frac{\rho_0}{A}\frac{\partial p}{\partial t}\left(1 - \frac{B}{A^2} p\right)$$

$$\nabla . \boldsymbol{u} = -\frac{1}{A}\frac{\partial p}{\partial t} + \left[\frac{B}{A^3} p \frac{\partial p}{\partial t} - \frac{1}{A} \nabla . (p\boldsymbol{u})\right]$$

Making use of the identity

$$\nabla . (p\boldsymbol{u}) = p \nabla . \boldsymbol{u} + \boldsymbol{u} \nabla p$$

we get

$$\nabla . \boldsymbol{u} = -\frac{1}{A}\frac{\partial p}{\partial t} + \left[\frac{B}{A^3} p \frac{\partial p}{\partial t} - \frac{1}{A} p \nabla . \boldsymbol{u} - \frac{1}{A} \boldsymbol{u} \nabla p\right]$$

Within the second-order expression enclosed in the square brackets, negligible error will be introduced by making use of the linear relationships

$$\nabla p = -\rho_0 \frac{\partial}{\partial t} \boldsymbol{u}$$

$$\nabla . \boldsymbol{u} = -\frac{1}{A}\frac{\partial p}{\partial t}$$

derived from Eqns. 6.9 and 6.10. This gives

$$\nabla . \boldsymbol{u} = -\frac{1}{A}\frac{\partial p}{\partial t} + \left[\frac{B}{A^3} p \frac{\partial p}{\partial t} + \frac{1}{A^2} p \frac{\partial p}{\partial t} + \frac{\rho_0}{A} \boldsymbol{u} \frac{\partial}{\partial t} \boldsymbol{u}\right]$$

$$= -\frac{1}{A}\frac{\partial p}{\partial t} + \frac{1}{2A}\frac{\partial}{\partial t}\left[\left(1 + \frac{B}{A}\right)\frac{p^2}{A} + \rho_0 u^2\right] \tag{6A.4}$$

We have now derived second-order approximations to the two basic equations (6.9 and 6.10); they are

$$\nabla p = -\rho_0 \frac{\partial}{\partial t} \boldsymbol{u} + \frac{1}{2} \nabla \left[\frac{p^2}{A} - \rho_0 u^2\right] \tag{6A.3}$$

$$\nabla . \boldsymbol{u} = -\frac{1}{A}\frac{\partial p}{\partial t} + \frac{1}{2A}\frac{\partial}{\partial t}\left[\left(1 + \frac{B}{A}\right)\frac{p^2}{A} + \rho_0 u^2\right] \tag{6A.4}$$

The next step is to derive the wave equation by combining these. This is done by taking the divergence of the first and differentiating the second with respect to time. The term $(\partial/\partial t)(\nabla . \boldsymbol{u})$ can then be eliminated from these two simultaneous equations to give

$$\nabla^2 p = \frac{\rho_0}{A}\frac{\partial^2 p}{\partial t^2} + \frac{1}{2} \nabla^2 \left[\frac{p^2}{A} - \rho_0 u^2\right] - \frac{\rho_0}{2A}\frac{\partial^2}{\partial t^2}\left[\left(1 + \frac{B}{A}\right)\frac{p^2}{A} + \rho_0 u^2\right] \tag{6A.5}$$

The two second-order terms in Eqn. 6A.5 can be combined by again making use of a linear approximation in order to rearrange a second-order term. In this case the assumption is that both p^2 and u^2 must obey linear wave equations so that

$$\nabla^2 p^2 = \frac{\rho_0}{A} \frac{\partial^2}{\partial t^2} p^2$$

and

$$\nabla^2 u^2 = \frac{\rho_0}{A} \frac{\partial^2}{\partial t^2} u^2$$

whence

$$\nabla\left[\frac{p^2}{A} - \rho_0 u^2\right] = \frac{\rho_0}{A} \frac{\partial^2}{\partial t^2}\left[\frac{p^2}{A} - \rho_0 u^2\right]$$

and Eqn. 6A.5 becomes

$$\nabla^2 p = \frac{\rho_0}{A} \frac{\partial^2 p}{\partial t^2} - \frac{\rho_0}{A} \frac{\partial^2}{\partial t^2}\left[\frac{B}{2A}\frac{p^2}{A} - \rho_0 u^2\right] \tag{6A.6}$$

$$= \frac{1}{C_0^2} \frac{\partial^2}{\partial t^2}\left[p - \frac{1}{A}\left(\frac{B}{2A} p^2 + Z_0 u^2\right)\right] \tag{6A.7}$$

It should be mentioned that there are more direct methods of deriving Eqn. 6A.7 without explicitly stating Eqns. 6A.3 and 6A.4. The present method has however been chosen deliberately with the intention of indicating more clearly the physical processes which contribute to the total non-linear term is the wave equation and also of removing any possibility of misunderstanding about the frame of reference used.

Appendix II

Source density for plane waves in different directions

Eqn. 6.26 applies generally to acoustic fields of any configuration, the only stipulation being that the particle velocity field must be irrotational.

A particular case which is of interest is that in which the primary field is obtained by superposing two uniform plane waves with different frequencies and different directions of propagation.

Suppose that the coordinate axes are chosen so that the two directions of propagation lie in the xy-plane and make angles θ_1 and θ_2 respectively with the positive x-axis. The pressure at any point can then be expressed as

$$p(x, y, t) = p_1 \cos M + p_2 \cos N$$

where p_1 and p_2 are the respective pressure amplitudes of the two waves and

$$M = \frac{\omega_1}{C_0}[(x \cos \theta_1 + y \sin \theta_1) - C_0 t]$$

$$N = \frac{\omega_2}{C_0}[(x \cos \theta_2 + y \sin \theta_2) - C_0 t]$$

The components of the velocity vector will be

$$Z_0u_x = p_1 \cos M \cos \theta_1 + p_2 \cos N \cos \theta_2$$
$$Z_0u_y = p_1 \cos M \sin \theta_1 + p_2 \cos N \sin \theta_2$$
$$u_z = 0$$

so that

$$(Z_0u)^2 = Z_0^2(u_x^2 + u_y^2)$$
$$= p_1^2 \cos^2 M + p_2^2 \cos^2 N + 2p_1p_2 \cos M \cos N \cos(\theta_1 - \theta_2)$$
$$p^2 = p_1^2 \cos^2 M + p_2^2 \cos^2 N + 2p_1p_2 \cos M \cos N$$

We can recognize two kinds of terms in these expressions. There are 'harmonic' terms proportional to p_1^2 or p_2^2 and 'interaction' terms proportional to p_1p_2. As far as the former are concerned $(Z_0u)^2$ is equivalent to p^2 while, for the interaction terms $(Z_0u)^2$ is equivalent to $p^2 \cos \theta$ where $\theta = (\theta_1 - \theta_2)$. Thus for a pair of uniform plane waves with an angle θ between their directions of propagation, the source densities are

$$q_{(\text{harmonic})} = \frac{1}{\rho_0^2 C_0^4}\left(1 + \frac{B}{2A}\right)\frac{\partial}{\partial t}p^2 \quad (6A.8)$$

$$q_{(\text{interaction})} = \frac{1}{\rho_0^2 C_0^4}\left(\cos\theta + \frac{B}{2A}\right)\frac{\partial}{\partial t}p^2 \quad (6A.9)$$

It will be noted that, as pointed out in Section 6.1.5, these two expressions will coincide if, and only if, the angle θ is zero. It is only in this special case that Eqn. 6A.8 correctly represents all the components produced.

Note, incidentally, that the interaction source density for a pair of plane waves is always proportional to $(\partial/\partial t)p^2$; the angle θ between their directions merely determines the value of the numerical factor $(\cos \theta + B/2A)$.

References

1 Westervelt, P. J., 'Parametric Acoustic Array', *J. Acoust. Soc. Am.* **35**, 535 (1963).

2 Berktay, H. O., 'Possible Exploitation of Non-linear Acoustics in Underwater Transmitting Applications', *J. Sound Vib.* **2**, 435 (1965).

3 Berktay, H. O., 'Parametric Amplification by the Use of Acoustic Non-linearities and some Possible Applications', Electronic and Electrical Engineering Dept. *Memorandum No. 228*, University of Birmingham (1965).

4 Dunn, D. J., M. Kuljis and V. G. Welsby, 'Non-linear Effects in a Focused Underwater Standing-wave Acoustic System', *J. Sound Vib.* **2**, 471 (1964).

5 Welsby, V. G., and M. H. Safar, 'Acoustic non-linearity due to micro-bubbles in water', *Acustica*, **22**, No. 3, 177 (1969/70).

7

Underwater Instrumentation

D. M. J. P. Manley
Sheldon's Farmhouse, Hook, Hampshire

7.1 Underwater transmitters

The seismic type of transmitter is used to obtain information concerning the nature of the strata immediately below the ocean bed and has been employed by oil companies for many years. The essential technique has been to detonate a large explosive charge close to the water surface so that the primary and surface-reflected waves are nearly coincident. However, because of the effects of cavitation and the water droplets due to the incidence of shock waves at the surface, it is found in practice that this explosive sound source cannot be regarded as a localized point-source explosion.

Impulsive sources give a short large-amplitude transient and in the case of explosives containing an active charge of TNT the waveform has an initial steep-front followed by an approximately exponential decay and a sequence of bubble pulses. Although the first bubble pulse has a total energy of the order of the steep fronted part of the waveform, it has in general a lower amplitude and the effect of the succeeding bubble pulses may be neglected. Finite-amplitude wave effects will result from the shock front but for points immediately removed from the neighbourhood of the explosion source and other disturbances, the pressure amplitude of the shock is given approximately by the empirical formula

$$P_0 = 3{\cdot}9 \times 10^8 (W^{1/3}/R)^{1{\cdot}13} \text{ dyne cm}^{-2}$$

where R is the range of the explosive in metres and W is the mass (lb) of TNT. The decay time T of the shock wavefront, i.e., the time to fall to $(1/e)$ of the initial amplitude P_0, is given by

$$T = 74 W^{1/3} (R/W^{1/3})^{0{\cdot}22} \ \mu\text{s}$$

and the time to the amplitude of the first bubble is

$$\tau = 13 W^{1/3} (z+10)^{5/6} \ \mu\text{s}$$

where z is depth (metres) of the detonated explosive.

For resolution of the signal a seismic pulse should have a short duration time and to overcome noise it should possess a large energy content. Multi-channel stacking is one method for obtaining a larger signal/noise ratio for a given pulse width but it demands reproducibility of the generated pulse shape. Considerable effort has therefore been directed towards the development of impulsive sources of controlled pulse shape using explosive mixtures of gases, exploding wires, compressed air and underwater sparks. However, as with the underwater explosion, these devices are also characterized by secondary oscillations which severely limit the resolution obtainable. Several methods have been developed in underwater spark systems with the objective of eliminating these secondary oscillations.

An alternative to short-duration single pulses of large peak amplitude are small-amplitude but long-duration oscillatory wave trains, to obtain greater signal energy, as used in radar and sonar; the received signals are effectively time-compressed. For seismic profiling the stacking system is simpler provided that a sufficient number of pulses can be transmitted relative to the boat speed.

The sparker sound source has been developed to give a shorter low-frequency pulse than the explosive one, and it can also be controlled more easily as regards timing and duration. The shorter duration, arising from the smaller gas cavities produced by ionization of the water, makes this method most suitable for shallow-water seismic arrays.

Fig. 7.1 shows a simple sparker discharge circuit, the operating voltages being of the order of 10 kV. These sparker discharge devices can handle

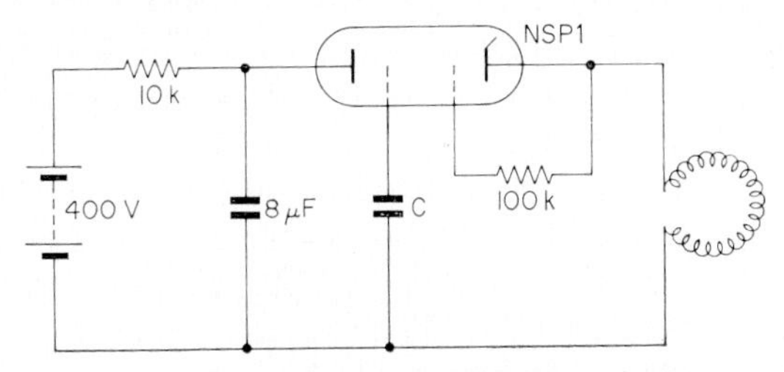

Fig. 7.1 A simple sparker discharge circuit

acoustic energies between 100 and 10,000 joules, with an acoustic efficiency of about 10 per cent. A representative receiver system would incorporate the receiver in a hydrophone array which is towed at 50–100 ft astern of the survey vessel. The electrical system should also include a harmonic filter to obtain a good signal/noise characteristic. Typical shallow seismic profiles on an experimental trace are given in Fig. 7.2. It should be mentioned that a good combustion gas explosion may be obtained using a mixture of propane and oxygen in a rubber bag containing an ignition plug. This device is suitable for work at greater depths, and has chief application in petroleum source investigations.

The thumper or boomer system is an alternative system which can be used to produce a low-frequency sound pulse and a typical electrical circuit is shown in simple detail in Fig. 7.3. The sound pulse results from the movement of a spring-loaded aluminium disc held close to a flat helical coil embedded in epoxy resin. The maximum effect is obtained by the discharge of a bank of low-inductance capacitors, charged from a high voltage source, giving a pulse duration time of a few milliseconds or less. Secondary oscillations in a thumper may arise from under-damped plate vibrations caused by cavitation, or by the ringing of the plate due to improper acoustic

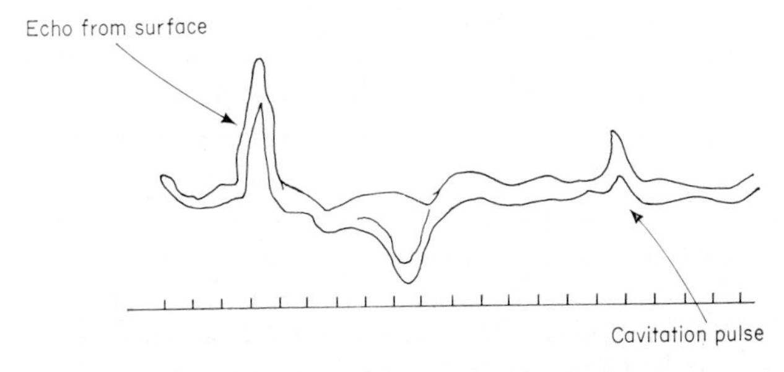

Fig. 7.2 Oscillogram of hydrophone output from the 1000 watt-second Boomer transducer signal. The negative sound pulse is due to the water surface reflection. Time-scale is one millisecond per division

matching to the water. Some thumpers are built on a purely hydraulic movement, but the electrical circuit has the advantage that a control can be kept on the exact firing point.

The acoustic spectrum from a thumper source has a discrete frequency maxima, due to the resonant frequency of the plate and associated gear. Depending on the area of cross-section of the plate, large thumpers can be designed to give energies up to 30,000 joules; their main disadvantage is the large mass of the system. Compared with the sparker, the thumper covers a lower-frequency spectrum and therefore achieves deeper propagation and larger ranges.

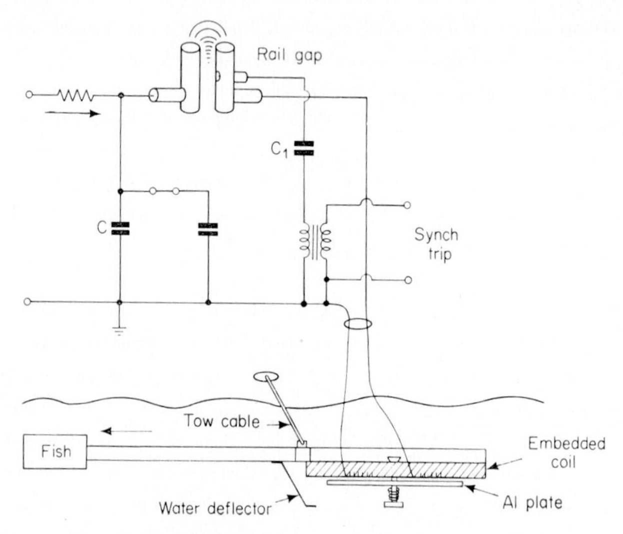

Fig. 7.3 Elementary circuit of the thumper unit

A variation of the spark method is the exploding-wire technique in which the discharge current passes through a fine wire connecting the sparking electrodes. Such a system produces higher acoustic pressures and lower-frequency spectra than obtained from a direct electrode discharge. Recent work by McGrath has shown that the performance is unaffected by the choice of wire material provided that the total energy involved in the discharge is from four to nine times the estimated vaporization energy of the wire material. At the lower-charging energies, for a given capacitor charging voltage, it is found that materials of low boiling points and heats of vaporization yield the highest peak pressures. The exploding-wire technique has particular application for deep off-shore profiling.

Hydrodynamic oscillators, which function by converting energy from a constant-pressure liquid source into an alternating pressure that is dissipated into the load, are especially advantageous for generating continuous high-level pure-tone signals at low frequencies.

7.1.1. *Magnetostrictive and piezoelectric transducers*

Acoustic transducers are devices for converting electrical or magnetic energy into the mechanical form, or *vice versa*, and usually refer to solid-state generators (or detectors) of ultrasonic radiation. The magnetic type depend for their action upon the change in length of a magnetostrictive material when subjected to an external magnetic field. Since the effect is independent of the direction of the field a static-biasing magnetic field is necessary for the length changes to keep in step with the alternations of the applied a.c. magnetic field.

The magnetic properties of magnetostrictive materials are structure sensitive, being dependent on the thermal and mechanical treatment they have undergone. Table 7.1 gives the average performance of typical materials. In practice the acoustical output of a magnetostrictive transducer is dependent not only on the value of the magnetostrictive coefficient (i.e., the strain $\Delta l/l$, where l is length), but also on the maximum stress it can sustain without fatigue failure and on its general design, both mechanical and electrical. The stresses that may be reached under dynamical conditions could easily approach the fatigue strength of nickel, i.e., of the order of 10,000 p.s.i.

Table 7.1 *Physical properties of piezomagnetic materials*

Material	Magneto-striction coefficient $\left(\frac{\Delta l}{l}\right)$ at saturation induction	Incremental magnetic permeability (optimum value) $H\,m^{-1}$	Electro-mechanical coupling coefficient k	Curie temperature °C	Young's modulus (Y) $10^{11}\,N\,m^{-2}$
Annealed nickel 99·92 Ni	-33×10^{-6}	$4\cdot3\times10^{-6}$	0·31	358	2·0
Permalloy 45 Ni : 55 Fe	$+27\times10^{-6}$	$2\cdot9\times10^{-4}$	0·12	440	1·4
Alfer 13 AZ : 87 Fe	$+40\times10^{-6}$	$2\cdot4\times10^{-4}$	0·27	500	1·5
Ferrous ferrite $Fe^{2+}+Fe_2^{3+}+O_4$	$+40\times10^{-6}$	$1\cdot9\times10^{-2}$	0·30	190	1·8
Ferrite 7 A2 Ni–Cu–Co ferrite	-28×10^{-6}	$6\cdot5\times10^{-5}$	0·23	640	1·7

A definite advantage of piezomagnetic over piezoelectric ceramics is the low transducer impedance, which may also be modified by varying the number of windings on the transducer. The lower impedance is particularly favourable for use with transistor systems.

Probably the most difficult problem in magnetostrictive oscillators is the induced eddy currents which reduce the effectiveness of the applied alternating field and so dissipate energy in the form of heat. By using thin insulated laminations and good design, vibrators which are radiating into liquids can attain conversion efficiencies of magnetic to acoustic energy up to 70 per cent. The upper practicable frequency of operation is about 100 kHz by reason of limitations due to increasing eddy currents losses and the fact that the plate form of vibrator is no longer possible.

The piezoelectric effect is concerned with the mechanical strains set up when certain crystals are placed in electric fields and is dependent upon the piezoelectric crystal having a structure which is to some extent non-symmetrical. Many crystals are piezoelectric but few combine this quality with sufficient mechanical strength and physical stability to be useful as transducers. By contrast with magnetostriction the sign of the stress changes with the direction of the applied electric field. Quartz is the most stable material and has wide application but for low frequencies quite prohibitively large crystals would be required. Hence recourse is directed towards composite systems comprising sandwiches of quartz crystals and steel plates. Other crystals used in underwater work include ADP (ammonium dihydrogen phosphate) which is chemically fairly stable and not markedly temperature dependent. It is somewhat hygroscopic so is used in an oil container which has an acoustically 'transparent' rubber diaphragm as a window. Tourmaline has piezoelectric coefficients similar to those of quartz but its internal losses are high. In contrast also to quartz, which is insensitive to hydrostatic pressure, tourmaline is sensitive to triaxial pressures applied uniformly around the crystal. Comparable with magnetostriction there exists in all dielectrics the phenomenon of electrostriction. The effect is normally extremely small but the advent of the titanate materials has led to their wide application as transducers, both as generators and detectors. After cooling from above its Curie point in a strong electric field (~ 10 to $20\ \mathrm{kV\ cm^{-1}}$) such a material will retain a permanent polarization for an indefinite period. They possess the obvious advantages of being easily cast into chosen shapes and of being polarized with any given field configuration. A form of composite ceramic–steel transducer is shown in Fig. 7.4. The two piezoelectric constants in common use are: the constant g which is the field produced in a piezoelectric crystal by an applied stress and is expressed in 10^{-3} metre volts per newton. The constant d, as defined with respect to the direct effect, is the electric charge density developed per unit applied mechanical stress and will be given in coulombs per newton. For the converse, or indirect, effect d is also the mechanical strain produced per unit applied electric field under no-load conditions and will be expressed as metres per volt. d will be numerically the same for both definitions.

g and d are directly related by the expression $d = K\epsilon_0 g$, where K is the dielectric constant of the material and ϵ_0, the permittivity of a vacuum, has

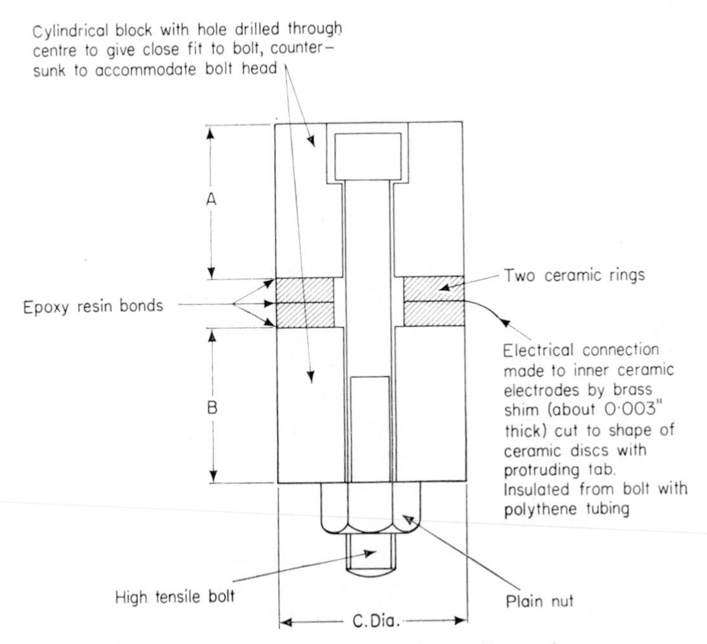

Fig. 7.4 A composite ceramic-steel transducer

the value 9×10^{-12} farad per metre. The coupling coefficient (k^2) which is a measure of the input electrical energy converted into mechanical energy, or *vice versa*, may be expressed as $k^2 = gdY$, where Y is Young's modulus of elasticity. It is evident that for optimum energy transfer both g and d should be high. For purposes of detection it is desirable that the material has a high g factor and furthermore increased sensitivity will result from a high dielectric constant, which will reduce possible loss in signal strength due to the capacitance of connecting cables which are effectively parallel with the transducer.

For polar ceramic materials the direction of positive polarization is generally taken to be that of the z axis. If the directions of the x, y and z axes are represented respectively by 1, 2 and 3 and the shear directions to these axes correspondingly by 4, 5 and 6, then the different related parameters may be written with subscripts referred to these. The first subscript is chosen to be the vector component of the electric field with the tensor component of the stress and strain as the second. For example d_{32} is the strain of a specimen in the y direction arising from an electric field applied along the z axis, or alternatively the electric charge density acquired in the z direction due to a mechanical stress exerted along the y axis. In the case of rectangular plates, discs and rings expanding along the axis of polarization the appropriate constants would be d_{33}, g_{33}, k_{33} and Y_{33}^{E}.

7.2 Underwater location devices

Only recently has the value of the underwater marking of sites become fully appreciated and oil companies have been satisfied generally with the use of simple omnidirectional acoustic beacons called 'pingers'. These have a pulse repetition frequency of from 1 per sec to 1 per 20 sec and consume about 10 W of electrical energy with an acoustic efficiency of about 10 per cent. Such devices have been produced with lifetimes up to 5 years and reliability over such long periods necessitates the use of mercury cells.

The National Institute of Oceanography have developed a method for beacon fixing in which a precision pinger is utilized. In this the repetition time constant is absolutely fixed and the device is used in conjunction with a facsimile recorder, or echo-sounder recorder, which has a drum or helix motor whose motion is controlled by a tuning fork or crystal controlled oscillator. The first reflected signals from the sea bed and sea surface are received by a hydrophone and recorded so that the position of the beacon may be deduced. There have been successful commercial approaches to directional receiver hydrophones which generally work on the principle of multi-elements of piezoelectric or magnetostrictive material. Such arrays are generally frequency dependent and are designed to match the pinger frequency.

There are, however, several other methods of scanning sonar devices that will either give accurate location to a precision pinger or in some cases give approximate location bearings on normal pingers. D. G. Tucker and others have used a linear array of receiving transducers and a tapped delay line. Several workers have published developments following this procedure.

The directivity control of the scanning sonar can be achieved by using the sum and the difference of several hydrophone outputs to control the sharpness of the directional pattern. This technique is known generally as the SD method.

The other method is the time-difference method which has been used by the author. In all cases of time-difference scanning, it is important to reshape the pulses received, which is best achieved by overloading the pulse signal and severely clipping the top to obtain as near as possible a square wave.

However, one important development in the time delay, or TD system, is the time delaying of the first received pulse to 'mash' or cover the second. This then constitutes a 'null' method of accurate assessment of the correct time delay because the combining of two similar pulses can be achieved very accurately and very small differences in any time delay between these two pulses can be detected.

7.2.1 Transducers as receivers

Underwater microphones, or hydrophones as they are called, are the sensors of the detecting system which includes preamplifiers, auxiliary circuits, transformers and cables. The transformers and preamplifiers serve to match the electrical impedance of the sensor to the cable, yet at the same time must possess the qualities of ruggedness and ability to withstand the problems set by the sea environment—high hydrostatic pressure, corrosion, etc. Although structural parts have been mostly fabricated in metal, particularly certain copper–nickel alloys, yet there is a breakthrough in the use of polyvinyl chloride and glass-impregnated resins as well as ceramics.

Sensors for hydrophones are now mostly piezoelectric sound-pressure devices and the common coupling modes employed are volume (*VE* or k_L), transverse (*LE* or k_{31}), longitudinal or thickness (*TE* or k_{33}) and planar (*PE* or k_p). Although the volume mode devices are unique in that all surface active elements may be exposed to the sound pressure and they are the most stable and reliable, yet the other modes are in more common use because of their higher sensitivities and lower self-noise characteristics. Moreover, by the modification of placing the sensor element in an oil-filled box so that the hydrostatic pressure is communicated via a sealing diaphragm the devices may be used in deep ocean.

Since normal piezoelectric hydrophones are equivalent electrically to capacitances with shunt resistances of the order of 5×10^{10} Ω they are not usually suitable for direct connection to more than two or three hundred feet of cable. On grounds of economy transformers may be used for electrical-impedance matching but now for most applications, outside of the lower sonic frequencies, the most commonly used devices are low-noise field-effect-transistor (FET) preamplifiers. Electrostatic shielding of the sensing element from external electrical fields, i.e., due to the power systems on a ship or otherwise, necessitates care in design to minimize ground-loop noise. In detecting explosive sounds there is the danger of the sensor system being overloaded and limiting devices should be incorporated in the circuitry.

The advent of the FET preamplifier has extended the lower frequency limit to the order of 0·01 Hz while the upper frequency rating is normally about 200 kHz.

If a hydrophone is to be omnidirectional then its maximum dimension will be determined by the wavelength of the sound to be detected. The sensitivity of detection is determined by the electrical self-noise level arising in the first stage of electronic amplification and the sound pressure level (SPL) that would produce the same level of electrical signal as the self-noise is termed 'the equivalent input noise sound pressure level' (ENSPL), which is expressed in dB relative to 1 μbar pressure. The ENSPL is directly related to the product of the electrical noise (in volts) at the input of the preamplifier and the sensor sensitivity (in dB relative to 1 V per μbar). For a given preamplifier the self-noise level will depend upon the nature of the piezoelectric sensor material, the particular coupling mode used and the number of electrodes involved. Of the various ceramics lead zirconates (PZT 5 and 4) are least noisy both in the longitudinal and planar modes. Rochelle salt (45° X-cut) in the transverse and lead metaniobate in the volume mode. The latter material is a new development and has a sufficiently large volume sensitivity to be used in the volume mode at great depths, i.e., greater than 3000 ft (i.e., ~1300 p.s.i.). Some sensors have been designed for operating at 35,000 ft. Changes of pressure and temperature can affect the stability of the sensitivity of the hydrophone, being dependent on the piezoelectric material and in the particular mode employed, and in this latter respect the most stable is the volume mode followed by the longitudinal, transverse and planar in that sequence. Lead metaniobate ceramic is the most favourable material and then come the lead zirconates 4 and 5. It should be added that over the normal working conditions the various crystalline piezoelectric materials such as quartz, lithium sulphate and ammonium dihydrogen phosphate are extremely stable. Sensitivities are expressed in terms of the *g* coefficient, which is the ratio of Electric field developed/Applied mechanical stress and expressed in M.K.S. units is V m N^{-1}.

In designing an underwater transducer to obtain the desired radiation characteristics it has been usual to incorporate some form of pressure-release material, such as an air-filled cellular material or an air cavity inside the housing of the transducer. The latter procedure involves a heavy housing to withstand the pressures likely to be experienced at great depth, while the cellular material would inevitably collapse. Efficient radiation characteristics over a wide frequency band have been obtained using short open cylindrical tubes of barium titanate or lead zirconate ceramic. The radial motion of the walls of a cylinder can excite the symmetrical cavity modes of the enclosed

Table 7.2 *Physical properties of some ferroelectric materials*

	Acoustic impedance 10^6 kg m^{-2} s^{-1}	Dielectric constant	Electro-mechanical coupling factor (k_{33}) thickness mode	Piezoelectric constant (d_{33}) thickness mode 10^{-12} mV^{-1}	Piezoelectric constant (g_{33}) thickness mode 10^{-3} Vm N^{-1}	Curie temperature °C	Young's modulus (Y) 10^{10} N m^{-2}
Quartz X-cut	15	4·5	0·1	2·3	58	575	8·0
Barium titanate ($BaTiO_3$)	24	1700	0·48	150	14	115	12
Lead-zirconate–titanate ($PbZiO_3 : PbTiO_3$) PZT_4	30	1300	0·64	280	26	320	8·2
Lead-zirconate–titanate ($PbZiO_3 : PbTiO_3$) PZT_5	28	1700	0·675	370	25	360	6·8
Lead metaniobate	16	225	0·07	85	42	550	2·9

column of water and this coupling obtained with short cylinders can give a high electroacoustic efficiency approximately over an octave frequency band. Furthermore, by using a coaxial arrangement of such cylindrical elements the directivity can be improved. The acoustic power output attainable for a typical system at about 20 kHz could be about 300 W, assuming that the ambient pressure prevents cavitation and the element is hoop prestressed to avoid mechanical failure due to the tensile strength of the ceramic being exceeded. The electric driving field is limited to about 3 kV cm^{-1} for lead zirconate to avoid increasing dielectric loss at higher frequencies.

7.2.2 *The vibrating-wire pressure transducer*

This is now a reliable and fully developed instrument for pressure measurements in oceanography. It effectively converts pressure to frequency by means of a taut wire which is attached to a pressure diaphragm. An alternating current at the natural frequency of the taut wire is passed along the wire, which is situated in the normally directed magnetic field of one or two permanent magnets suitably arranged. Under the mutual action of magnetic field and electric current the wire will perform transverse vibrations at a frequency dependent on the tautness of the wire, which will be modified by the movement of the diaphragm. In other words the natural frequency of vibration is directly dependent on the pressure applied to the diaphragm. To avoid any influence of the vibrating wire upon the diaphragm (which has a controlled thermoelastic coefficient) stiffness the wire is supported on a frame independent of the diaphragm, the materials of the frame being chosen to give a thermal expansion coefficient to negate any change of length of the wire with temperature. It should be mentioned that the mechanics of the vibrating wire can become quite complicated in the resonant condition when the amplitude of vibration increases, leading to the jump phenomena and to a whirling motion. Hence for stable working the drive current must be below a certain determined value. In practice a typical system will consist, below the sea surface, of the transducer with an exciter amplifier, a transmitter buffer amplifier, a suitable d.c. power supply and a temperature chamber with its heater jacket and controller. A suitable cable connects this system with a power controller and receiver buffer above the surface, and the transducer has been used to measure pressures at depths of over 20,000 ft.

7.3 Basic telemetering systems

In order to carry information under water without use of cables, some system has to be used in order to modulate a carrier of information through the water or over the sea surface. Possible wave systems could be electromagnetic, optical and acoustical. For example, the sea surface forms an excellent medium over which radio information may be transmitted with the minimum of interference while optical laser beams have potential applications in clear media.

If acoustic radiation is propagated under water at frequencies below 100 kHz, attentuation is not a disturbing factor, but one of the main problems is the effect of the multiple interference phenomena on modulated information. A statistical study has been made of the random fluctuations arising from these phenomena and for short ranges the fluctuations obey a three-halves power law, and become constant in amplitude for large ranges, although the phase variations increase linearly with distance. The very variable contour of the sea bottom, in shallow waters, leads to a worsening of the operation.

In ideal conditions the factors limiting underwater telemetering are:

1. Transmitter power
2. The gain of directional transmitting and receiving arrays
3. 'Spread' of the propagated signal
4. Absorption of the signal by water
5. Noise experienced in the receiver and associated circuits.

A number of recent developments commercially has shown that by careful transformer design the conventional PZT4 piezoelectric and the corresponding heavy metal magnetostrictive, transducers work efficiently with little 'ringing' effects. The transducer circuit must be critically damped, however. The practical limitation to the amount of power is the point of cavitation inception on the transducer face and before this point is reached, in some cases, much aeration can occur. About 20–40 W acoustic power can be emitted from a transducer face of 100 cm^2 before the point of aeration is reached at moderate depths.

The techniques employed in the transmitting array are fundamentally simple, relying on phase support of the acoustic signal. More sophisticated devices can incorporate electronic control of the 'signal phase' rather than on the variation of the physical spacing. Typical arrays would have a beamwidth of 15° between the half-power points and the resultant transmitter directional gain would be of the order of 200. Similar techniques are used for the receiver array and, if both transmitter and receiver are directional, then difficulties occur over alignment which has to be performed experimentally in view of refraction of the acoustic rays.

The problem of receiver noise arises from a number of causes such as motion of sea water over the transducer face and 'sea-state' noise due to waves and surface layer activity. To minimize the effects a narrow bandwidth should be used, and in telemetering work this is best achieved by high-pass and low-pass filters.

In order to obtain information bandwidth in the transducers as high a frequency is needed as possible, but against this the theoretical range diminishes as the frequency is increased. Long ranges in excess of 10 miles are only possible at frequencies in the audible range and below.

Multi-path propagation does limit the rate of transmission of information, and if a simple pulse is fed into multi-path propagation both the width of the pulse and decay-time are lengthened. For a simple sinusoidal waveform, multi-path effects causes random fluctuation of the phase of the received signals.

Diversity techniques can be used to offset multi-channel fading. In a simple diversity system, several receivers or tuned transducers of different frequencies are used. Automatic selection is made of the greatest amplitude signal so that any Doppler shift due to multi-channel effects is automatically compensated by the multi-frequency channels that can be chosen, so that this type of fading or interference is reduced.

Digital pulse-compression techniques can be used to offset frequency-selective fading due to multi-path effects. Binary signals can then be fed to modulate a carrier whose frequency is being varied to cover the frequency dispersion. If the carrier frequency is much greater than the binary information signal, then this technique, although covering the frequency diversity, will not affect the basic binary information.

Much work has now been done in order to improve information content in binary form still keeping within a limit in which distortion will not destroy or merge pulse information.

In feeding information to craft travelling at normal speeds, Doppler shifts would cause signal errors of an appreciable amount. One successful avoidance method is to use self-tracking filters, e.g., filters that can alter their bandwidth to suit the relative cruising conditions. In using a self-tracking system on to moving craft the frequency of the transmitter as well as receiver filters are off-set. Obviously, the direction of the craft does affect this procedure, but one can operate on a guidance carrier whose frequency shift will control both Doppler shifting as well as phase changes.

7.4 Current and wave measurements, monitoring of tides

An important aspect of underwater acoustic instrumentation concerns the devices employed to monitor water velocity and short and long term changes in the surface level of rivers or of the sea.

7.4.1 Current meters

Although not primarily an acoustic device the simple impeller type of instrument utilizes procedures applicable to the handling of acoustical data. The impeller will rotate at a speed approximately equal to the current velocity, and by the use of sensing coils mounted in close proximity to the impeller shaft, digital pulses are generated proportional to the revolution rate of turning. Reasonable speed stability may be obtained and time marking added, utilizing a crystal-controlled clock. The impeller technique gives large errors when measuring fast flows, or when turbulence is present, and the use of Doppler devices is to be preferred under these circumstances.

In the Doppler technique a continuous ultrasonic beam is propagated over a path length of approximately 60 cm at a frequency of 10 MHz. Two similar paths, using two pairs of similar transducers, are set up adjacent to each other. The direction of propagation is reversed between each path and current flow, depending on its direction, will cause an increase or decrease of frequency. By mixing the signals received from the two paths the resulting Doppler frequency shift will be proportional to twice the current velocity and digital-type output pulses may be produced, using a suitable pulse control circuit.

7.4.2 Wave and tide recording

Acoustic-type devices have received application recently to the monitoring of tides and waves, and one method employs the standard echo-sounding technique with the transducer probes directed upwards from the sea bed or underwater platform. To obtain the necessary narrow beam angle a high frequency of transmission (which will incur a large attenuation) is required, and a high pulse transmission rate is essential.

The information obtained can give direct wave profiles by use of a d.c. ramp generator, or else FM signals could be produced by the use of a crystal-controlled generator. There are sophisticated techniques by which the time path can be automatically backed-off when transducer probes are deep down from the waves. One main handicap to this method is the occasional presence of air bubbles near the surface waves, and these will act as good scattering and reflecting layers and thus lead to error signals.

Another measuring system makes use of a large plate capacitor of approximately 1 ft^2 in area and having an evacuated interspace between the plates. Changes of hydrostatic pressure will deflect the upper plate so changing the capacitance which forms a part of an oscillatory circuit. Changes of the order of 0·01 picofarads may be detected using suitable electronic circuitry which should be in close proximity to the capacitor to minimize any stray 'pick-up' effects. Pressure variations of the order of 0·001 in of water are detectable by this capacitance technique. However, it is not likely that any high-frequency components of the wave spectrum will be detectable below about 10 ft of water owing to their large attenuation. If the oscillator is beat with a standard oscillator the information will be presented as an audio-frequency modulated signal which may fulfil the function of a FM system of a low speed tape recorder.

An alternative technique to the capacitance and echo-detecting systems involves the use of an accelerometer to monitor the motions of a light buoy at the sea surface. The displacement spectrum may be derived from the recorded acceleration spectrum by use of a double integrating circuit.

7.5 Acoustic navigation systems

The development of these systems as precise navigational methods results from intensive work on electroacoustic transducers, electronic circuits, power supplies and supporting structures as regards their capability of operating over long periods of time in the deep ocean. The general principle of operation is for a pulsed command or interrogating signal to be transmitted from an acoustic transducer mounted below the ship's hull or in tow. The signal on being received by the receiver transducer is filtered and amplified by suitable electrical circuits and in turn a pulsed acoustical reply is triggered at a different frequency (or frequencies). The time interval between sending and receiving signals at the ship is convertible to range with an accuracy dependent on the accuracy of the known value of the average velocity of sound over the path. Assuming a 0·1 per cent accuracy it will mean that the error in a single range measurement at 15 miles is approximately 30 yd; frequencies up to about 12 kHz may be used. Higher frequencies are possible for shorter operating ranges, and coding of signals may be used for identification or security purposes. The unlimited depth capability and the long life of acoustic transducers suggests their use as shipping-lane markers with reference points situated say at every 200 miles. Their cost will depend on frequency, power output and length of life: the lower the frequency then the larger the transducer and the larger the floats required and hence increased cost.

Doppler sonar is similar to the Doppler radar techniques and the development of the system has provided excellent navigational tools for large tankers, cargo vessels, etc. on the one hand, and deep submergence vehicles on the other. The basis of operation is the use of four sonar beams which are respectively directed forward, aft, port and starboard, and the velocity is determined relative to the ocean floor by measuring the Doppler shift of each beam. In this way the ship's velocity along each of the axes is determinable and the drift angles may be computed, as may the velocity components displayed along the desired coordinates. With this information it becomes possible to navigate the ship from any starting position to any given terminal location by the coupling of the Doppler system to a suitable heading reference.

The early application of Doppler sonar navigation was restricted to shallow water due to the limitations of transducer and electronics technology. In recent years pulse Doppler systems have been successfully developed leading to a much wider application. The increasing size of tankers with their inherent large mass has emphasized the importance of accurately estimating the approach velocity of the ship towards the docking stage, and in such cases Doppler sonar has an important application.

7.5.1 *Advanced echo-sounding equipment*

Some examples of sophisticated equipment used under water are described briefly below and the first is the narrow-beam echo-ranger which is now a commercial marketed device. Fig. 7.5 shows a block diagram of the device which is a complex 'asdic' system and it provides a panoramic map of the ocean bed as well as being used for fish detection. The acoustic beam system is 'fan' shaped, the main beam having sufficient directivity for the 3 dB points to be $\pm 1{\cdot}3°$.

The transducer unit consists of three rows of nine sections each, each section consisting of 32 magnetostrictive elements made of Permalloy

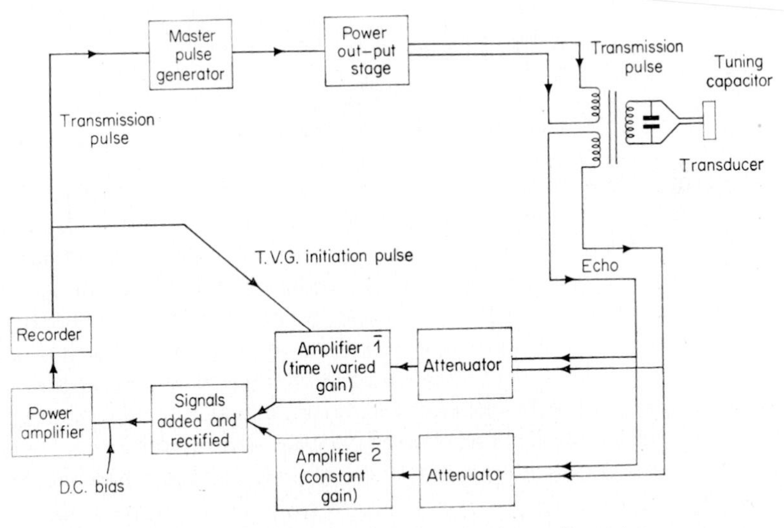

Fig. 7.5 Block diagram of a complex asdic system

stampings. The active face of the complete unit is 156 cm × 31·5 cm. A fibreglass window covers the sections in order to provide protection, and the complete unit is housed in a bronze casting.

Such a unit described is both expensive and heavy, but the unit can be rotated about an axis and the angle adjusted by a simple control device. The unit is mounted on the hull of the ship or can be specially towed just below the surface.

Fig. 7.6 shows the horizontal beam pattern, and Figs. 7.7(a) and (b) show the vertical beam pattern where 3 lobes are shown as to be expected

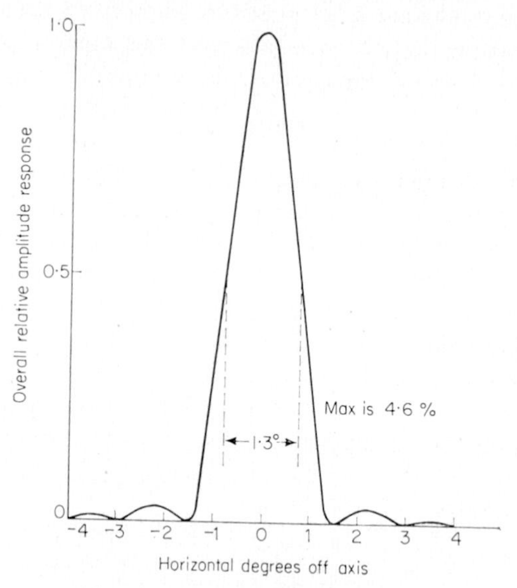

Fig. 7.6 Horizontal beam pattern

from normal diffraction theory. These lobes can be radically changed in importance by using only one or two rows of elements instead of all three.

The first of these units was designed to work at 36 kHz—the choice of frequency being a compromise between weight and cost of the transducer (this increases rapidly the lower the frequency considered) and the penetration required, for absorption falls rapidly with frequency. Basically the transmitter is triggered from contacts on a facsimile recorder. The transmitter is a typical valve or transistorized blocking oscillator stage, although in using a transistor design it is necessary to have an inverter stage or stages in order to produce high enough voltage to drive enough power into the transducer.

The receiver consists of a tuned voltage amplifier with a variable gain

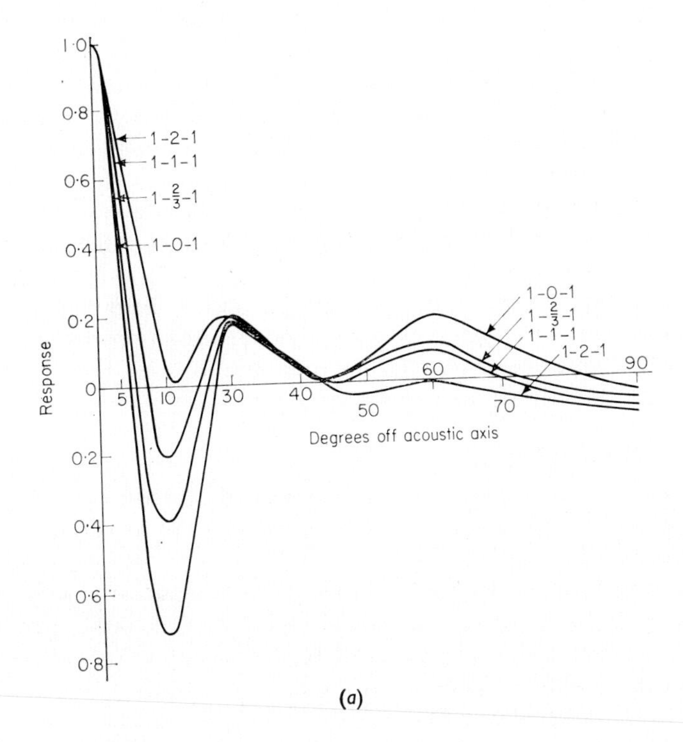

(a)

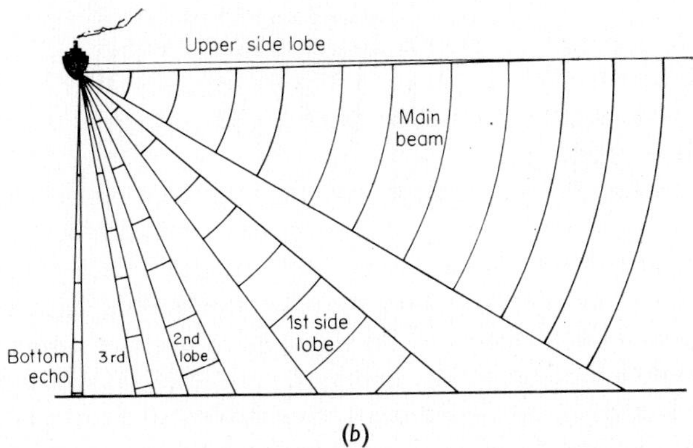

(b)

Fig. 7.7 (a) Vertical beam pattern. (b) Diagrammatic arrangement of the pattern

control so that the gain of the amplifier is increased during the sweep time of the recorder. This is in order to increase the weaker more distant signals to give uniform signal sensitivity.

Many manufacturers have incorporated the Muirhead 'Mufax' picture recorder working on the basis of producing pictures from telegraphic information. The basis of such a unit is the feed of high-voltage signals across a moving helix and knife edge. The helix turns at a rapid rate and the point of contact between the helix and the knife edge traces over the paper very rapidly. As a high-voltage signal appears, bleaching of the paper occurs and leaves a dark trace.

A recording technique of comparatively recent development utilizes ultraviolet light which has a possible disadvantage in operational conditions as it is not visible until some time after the recording. Fig. 7.8 shows a typical record given by the narrow-beam echo-sounder.

7.5.2 The mud probe

Another type of sophisticated development of the basic echo-sounder is the mud probe. The purpose of this device is to look at small thicknesses of silt or mud.

Because a muddy, silty sea-bed has an acoustic impedance which is not very different from that of water, some sound will penetrate a muddy sea-bed. Penetration will mainly occur at frequencies below 150 kHz, and the depth of penetration depends on two factors: (1) strength of the transmitted pulse, and (2) sensitivity of the receiver. In order to prevent effects of ringing in the receiver, and also in order to prevent loss of sensitivity, it is advisable to operate on a narrow bandpass filter of about 1 kHz either side of the probe frequency. The choice of this frequency is fairly important, for there has to be sufficient resolution to be able to discriminate very small thicknesses of mud.

One of the significant aspects of oceanographic instrumentation in recent years has been the introduction of modern electronic computers and data-processing equipment aboard survey and research vessels. Such systems have two basic modes of operation, the on-station and the underway. In the former case the vessel is stopped for survey purposes, the sensor is lowered at a maximum rate of 2 m s^{-1} to a depth of the order of 4000 m and the parameters measured are usually ambient temperature, sound velocity, salinity and depth. The average acoustic velocity is computed at the end of each 100 m depth interval. From bucket samples surface-water salinity and surface oxygen are determined and other data are calculated from Nausen charts, all computed data being recorded on magnetic tape. In underway systems gravity, total magnetic field, surface water parameters, etc. are

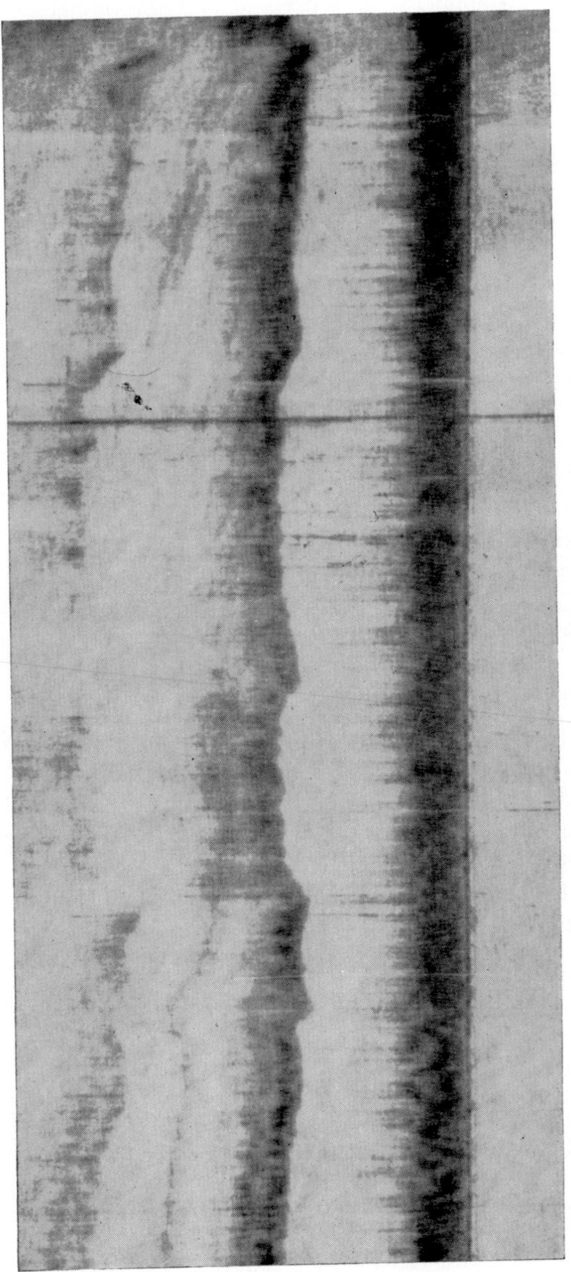

Fig. 7.8 Record from a narrow-beam echo-sounder

included. The computer will process data both on-line, continuously according to a fixed routine, and off-line at times when certain data become available. True-bottom depth is computed by means of an equation which allows for the variation of acoustic velocity with depth at depths greater than 2000 ft. The corrections may be made with sufficient accuracy to permit automatic machine correction if the average surface layer velocity is inserted into the computer at what need only be longish intervals, since generally any changes in velocity will be slow.

Looking to the future, work is continuing on the subject of acoustic imaging and with the study of underwater acoustic holography and these may yield interesting developments within the next 10 years or so. With the movement of the offshore oil industry towards deeper waters, the functions which have been performed by divers, cables and control lines are being replaced by acoustic control, telemetry and guidance systems. One such device is an acoustically controlled nuclear-powered blow-out preventer stack which obviates the need for control cables and hydraulic lines between the well-head and the drilling vessel. The system has operated in several hundred feet of water and the acoustic link is broken when a blow-out occurs, and the drill pipe is automatically sheared-off and the hole is closed, so ensuring the safety of the crew of the drilling vessel. Mention has been made earlier of acoustic beacons and these are now being utilized for visual displaying of the position of a coring vessel with respect to the core hole. The acoustic signal is processed by a special-type of computer whose output drives high-powered outboard engines which hold the vessel precisely in position.

The use of acoustic techniques to 'home-in' on an established well-head has also found application in a number of methods. In one system, used at depths of 1000 ft, the slant range to the well is utilized to fix its rough location and when the drill pipe is directly over the well an additional acoustic signal is obtained. The display is on a double-beam oscilloscope.

Bibliography

Underwater transmitters

Weston, D. E., 'Underwater Explosions as Acoustic Sources', *Proc. Phys. Soc. Lond.* **76**, 233 (1960).

Wood, A. B., 'Model Experiments on Propagation in Shallow Seas', *J. Acoust. Soc. Am.* **31**, 9, 1213 (1959).

Lecort, M. D., 'Vibrating Wire Pressure Transducer Technology', *J. Ocean Technol.* **2**, 37 (1968).

Hutchins, R. W., 'Broadband Electroacoustic Sources for High Resolution Sub-bottom Profiling', *Oceanology International 69*, Part I, Feb. 1969.

Boyer, D. W., 'Spherical Explosions and Implosions', *D.R.B., U.S.N., U.T.I.A. Report*, No. 58.

Hydrodynamic oscillators

Bouyoucos, J. V., and F. V. Hunt, U.S. Patent 3,004,512.

Blackburn, J. F., G. Reethof, and J. L. Shearer, *Fluid Power Control*, Wiley, New York, 1960.

Liebich, R. E., *Oceanology International 69* **51**, May/June, 1969.

Transducers

Hunt, F. V., *Electroacoustics*, Harvard U.P., Cambridge, Mass., 1954.

Hueter, T. F., and R. H. Bolt, *Sonics*, Wiley, New York, 1955.

Van der Burgt, C. M., 'Piezomagnetic Ferrites', *Electron. Technol.*, Sept. 1960.

Van der Burgt, C. M., and H. S. J. Piyls, *Inst. Elec. Engrs Trans.* (Ultrasonic Engineering) **UE-10**, 1, July, 1963.

Frederick, D. R., *Ultrasonic Engineering*, Wiley, New York, 1965.

Schofield, D., 'Transducer Tables and Materials', *Underwater Acoustics Nato Conference* (V. M. Albers, ed.), Plenum, New York, 1961.

Sonar systems and telemetering

Hersey, J. B., H. E. Edgerton, S. O. Raymond, and G. Hayward, 'Sonar Uses in Oceanography', Instrument Society of America, *Conference Preprint* **21–60**, 5–9 (1960).

Lawrence, G., *Electronics in Oceanography*, Ch. 10, Foulsham, Slough, 1967.

Chesterman, W. D., P. R. Clynick, and A. H. Stride, 'An Acoustic Aid to Sea-bed Survey', *Acustica*, **8**, 285 (1958).

Tucker, D. G., 'Sonar Arrays, Systems and Displays', *Underwater Acoustics Nato Conference* (V. N. Albers, ed.), Plenum, New York, 1961.

Klein, M., 'Side-scan Sonar', *Undersea Technol.* **8**, 24 (1967).

Stephens, F. H., and F. J. Shea, 'Underwater Telemeter for Depth and Temperature', U.S. Fish Wildlife Serv. *Spec. Sci. Rep. Fisheries* **181**.

Caldwell, D. R., F. E. Snodgrass and M. H. Wimbush, 'Sensors in the Deep Sea', *Phys Today*, July 1969.

8

Audio Communication between Free Divers

B. Ray
Physics Department, Imperial College of Science and Technology, London

8.1 Introduction

Before looking at the communication problems encountered by men working under the sea, it is necessary to examine the equipment and techniques that are used. This introductory section will be devoted to a brief examination of diving technology.

Until the present decade nearly all serious underwater work was performed using a system in which the diver was connected to the surface at all times. The diver wears a 'Hard-Hat' diving dress, which comprises a heavy copper helmet attached to a rubber suit of generous proportions, together with the necessary lead weights and heavy lead boots to keep the diver stable on the sea floor. In Hard-Hat, as with other diving systems, the human body is exposed to the ambient water pressure; however if all the cavities within the human body are 'equalized' to this pressure then the diver may be quite unaware of the absolute pressure. Since the body will not tolerate a pressure differential, and if the diver has failed to equalize the pressure in his lungs, ears or sinuses by more than a few feet water gauge, the results can be very serious. It is imperative, therefore, in diving equipment to provide breathing gas at the ambient water pressure; in the Hard-Hat system this requirement is met by an air line to a surface compressor which continually supplies the diver with excess air. The correct air pressure is then maintained in the diving dress by expelling this air into the water through a valve on the helmet. Communication with the surface is maintained by a telephone cable alongside the air line. There is ample space inside the copper helmet to fit an intercom using a conventional moving-coil loudspeaker and microphone.

Since World War II compressed air breathing apparatus† has become available. The heart of a compressed-air diving set is the 'demand-valve'. This is usually a one or two stage pressure reducer attached to a high-pressure air cylinder. It is designed to provide air at a pressure that will remain within about 2 in water gauge of the pressure at its sensing diaphragm. Because of the sensitivity of the demand valve, it is necessary to return the exhaled air to a point close to the sensing diaphragm before releasing it through a non-return valve. The self-contained compressed-air diver may wear a mask covering his eyes and nose and hold his breathing hoses between his teeth with a mouth bit; he may wear a mask covering his eyes and nose and a second cup-shaped mask over his mouth; or he may use a 'full-face' mask covering the whole of the face. For protection against the cold a close-fitting rubber suit is normally worn and a pair of fins give the free diver far more mobility than his predecessors. The acoustic problems brought about by the use of a mask will be dealt with later. However, at this stage it should be noted that broad-band noise will be produced by the pressure reduction and this will completely mask most hearing. In the exhalation process there will be the operation of several non-return valves and the production and release of bubbles.

The equipment described so far is not suitable for work below about 250 ft. For deep diving it is usual to use some form of closed- or semi-closed-circuit breathing equipment. In this apparatus the breathing gas, which is now probably a mixture of oxygen and helium, is recirculated and used again. The exhaled gases pass over a chemical to remove carbon dioxide. A small bleed of high-pressure oxygen makes up for the amount of oxygen used by the human body, and the gases finally pass into a flexible bag. This bag is known as the counter-lung, from which inhalation takes place. Although the ancillary equipment tends to be similar to that of the compressed-air diver, the level of inhalation and exhalation noise is very much less.

† This is often called SCUBA in the U.S. and AQUA-LUNG in the U.K. These terms will not be used here as they are trade marks in certain countries.

8.2 The problem

The use of free divers and modern diving techniques however, has brought about some new problems with communication. The most important are listed below

(1) The free diver cannot use a cable telephone.

(2) The free diver's 'helmet' fits his head far closer than does the 'Hard-Hat', so creating problems in forming words into a confined, or non-existent cavity, and hearing a signal under similar conditions.

(3) The modern diver may well be using a breathing gas that differs markedly from air. The main reason for this is that air, or to be precise, the nitrogen in the air, has a narcotic effect under pressure. It is generally not practical to use oxygen/nitrogen breathing mixtures at depths below 250 ft and either helium or hydrogen are normally used as the inert gas for deeper diving; both helium and hydrogen have a velocity of sound that is greater than that in air.

In the context of these problems it is important not to take too much for granted. The communication engineer must consider all communication methods and must not reject such possibilities as hand signals. However it is proposed to limit this discussion to voice communication. In general this can be represented as in Fig. 8.1.

8.2.1 Before words are formed

The diver has many physiological enemies such as cold, anxiety, the build-up of exhaled carbon dioxide, anoxia due to inefficient breathing equipment, and narcosis; these may all have the effect, in the first place, of slowing down the mental processes of the diver. It is important that the communication equipment does not aggravate these troubles, nor must operating the equipment require considerable mental effort.

One interesting example of this last part is the use of 'mouth-masks', that is rubber cup-shaped mouldings designed to strap over the mouth and lips. Their object is to remove the requirement for the diver to hold a conventional 'bit' between his teeth. Although these devices do, in many cases, offer improved comfort and clearer communication, they can also cause problems in sealing the mask to the face. When a mouth-mask was used in a U.S. Navy saturation diving project[1] this problem was magnified to a dangerous degree by the back pressure exerted by the breathing apparatus. Needless to say, no intelligible communication was received from these divers!

The author has himself experienced conditions under which he was unable to remember which end of a vertical rope led to the surface; the reader is left to imagine how reliably one could operate a transmitter under these circumstances.

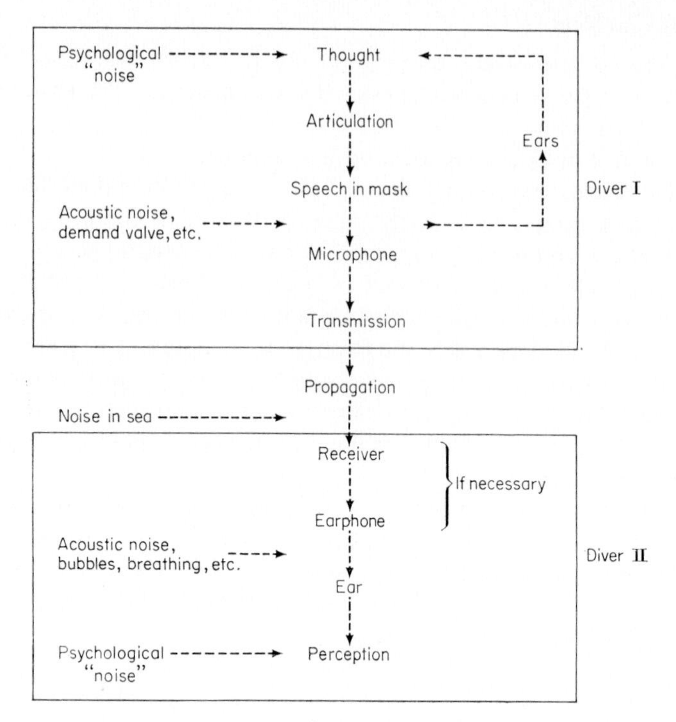

Fig. 8.1

8.2.2 *Word formation under water*

The main differences between speaking under water and speaking in air are shown in Table 8.1 which should be largely self-explanatory. The 'Contrary factors' column gives the reasons why the obvious solution may not be a suitable one. For example, one may try to overcome problems caused by the limited volume of the mask by increasing the volume, but this unfortunately can often give rise to pockets of expired air becoming trapped in the additional volume.

It is proposed to limit the discussion to two of the more important problems.

8.2.3 *Problems associated with breathing mixture*

Before discussing the effects of using breathing gases other than air, it may be useful to examine the reasons that drive physiologists to look for exotic mixtures. In general it is the partial pressure of the constituents of the

Table 8.1 *Comparison between speaking underwater and speaking in free air*

Difference	Effect	Contrary factors	Possible solutions
Mask has a finite volume	Acoustic impedance differs from air	Larger volume can cause problems with CO_2 build up	(a) Inner mask (b) Electronic feedback
The nose is generally blocked, or in a separate cavity	Uvula may be closed, giving rise to a change in vowel sounds	Requirement to 'clear' the ears	Mask that allows nose to be pinched for ear clearing, but with nose normally free in main cavity
The air path between mouth and ears does not exist (except with a helmet)	The feedback between voice and hearing is radically altered		(a) Helmet covering whole head (b) 'Side-tone' fed to ears
Water (or rubber) surrounding the chest and throat	Change in damping of the throat cavity and lack of radiation from throat or chest		
Mechanical restriction of mask on face muscles	Difficulty in producing certain speech sounds		Better masks
Increased velocity of sound in breathing gas when using helium–oxygen mixtures	'Donald Duck' effect on voice	At deeper depths air cannot be used (inert gas narcosis)	(a) Within limits it may be possible to add small amounts of nitrogen (b) Unscrambler
Increase in ambient pressure	Believed to cause a lack of fricatives and inability to whistle		
Small pressure fluctuations caused by breathing equipment. (Generally about 1 to 2 in water gauge)	'Difficulty' in articulation	Reliability of breathing equipment sets a limit on the sensitivity of the non-return valves used in the construction	Better design of breathing apparatus

inhaled gas that define its suitability for breathing. Oxygen is poisonous above 2 atmospheres partial pressure and nitrogen has a narcotic effect above about 7 atmospheres. Hence it should be clear that below about 200 ft it is necessary to reduce the percentage of oxygen and to use a substitute for nitrogen. The most common substitutes are hydrogen and helium. In mixtures containing either of these gases the sound velocity is greater than in air.†

The classical model for the human voice considers the larynx as a wide-band noise generator with the mouth, throat and nose cavities acting as resonators. The end result is a frequency spectrum whose envelope is defined by the vocal tract and whose fine structure is a function of the larynx waveform. The effect of the increased velocity of sound in helium/hydrogen mixtures is to shift the envelope shape up the frequency spectrum. The larynx pitch is not affected to a significant degree. The maxima of the frequency spectrum are termed the 'formants' of the voice and the most obvious effect of breathing a helium or hydrogen mixture is an increase in the frequency of these formants, the increase being in proportion to the change in the velocity of sound in the exhaled breath.

From the foregoing discussion it should be clear that in order to render helium speech intelligible it is pointless to reduce the component frequencies of speech by a fixed number of cycles per second (a relatively straight-forward operation). What is required is a proportional frequency reduction, which is just what happens when a tape recorder is played back at slow speed. Although under some circumstances it may be useful to use a tape recorder for this purpose, the distortion of the time scale normally negates any advantage gained. One practical method does, in fact, use a time-stretching technique to correct the frequency spectra. However a part of each separate sound is discarded so that the overall time scale remains true.[2] At the time of writing the most popular method of unscrambling the speech of divers is to break the speech signal into a number of frequency bands and to reduce the frequency of each by a different, fixed, amount. The larger the number of frequency bands the closer will be the approximation to the ideal.

Unfortunately, at present the performance of even the best instruments leaves a considerable amount to be desired. It appears that apart from shifting the formants, the gas mixture distorts speech in other ways. The role played by the absolute pressure is still in doubt; a mixture of 21 per cent oxygen, 79 per cent helium has the predicted effect at atmospheric pressure and can be unscrambled remarkably well. At 20 atmospheres pressure, a suitable breathing gas would contain less than 5 per cent oxygen and the

† The reduction of the oxygen percentage makes it possible to design a diving system that avoids explosive combinations of oxy-hydrogen.

velocity of sound in this would be higher than in the 21/79 mixture that was used at the surface. To the untrained observer the surface 21/79 speech would appear strange but understandable; however, the speech from 20 atmospheres would be completely unintelligible, and might not be recognized as the human voice.

The problems of making measurements under high pressures are many, for not only does there exist some doubt as to the nature of speech recorded under these conditions, but there are limitations on the gas mixtures that can be used in the case of human subjects. Furthermore, experiments at 20 atmospheres are very expensive to perform because it will take several days to decompress a subject from even a short exposure to this pressure.

8.2.4 Acoustic impedance of the facemask

If one listens to the signals received from a fully equipped diver who is standing with his breathing set submerged but with his head above the surface of the water, as he submerges his body completely, there will be a noticeable drop in intelligibility. In air there is appreciable transmission through the relatively thin walls of the facemask, which reduces the acoustic impedance of the mask cavity, as seen from the lips, below that of an infinitely rigid mask. When the mask is completely submerged, transmission through the walls is negligible and the mask can be regarded as a closed cavity having a high acoustic impedance. Although this argument strictly only holds for frequencies below the cavity resonance of the mask, the conclusion, namely that the mask will produce an effect which is inversely proportional to its volume (i.e., proportional to its impedance), appears to provide a useful guide to the choice of masks for practical communication purposes[3].

Two possible solutions to this problem will be considered. The obvious one is to increase the volume of the mask. If this is done, then some lightweight inner mask can be inserted to prevent a large dead-space as the latter leads to the possibility of carbon dioxide build-up. The second is to add an acoustic feedback path by placing a microphone, amplifier and loudspeaker inside the mask. It should be possible to design such a system to reduce the acoustic impedance seen from the lips. Although this technique has been used for sound absorption in air[4] there are practical difficulties in applying it underwater.

8.3 Microphones

Three types of microphone are in use by divers; they are air, bone-conduction and throat. All of them have particular problems. The air microphone, positioned in the mask, stands the best chance of receiving intelligible signals. However, it must be designed to withstand a large pressure change and frequent splashing with water. A design for a pressure-compensated air microphone is given in Fig. 8.2. By contrast the bone-conduction microphone is a vibration pick up[5] and can more readily be made as an encapsulated pressure-resistant device. Throat microphones[6] are not normally used underwater as the quality of reproduction is seldom sufficient to allow intelligible communication.

The choice between air and bone-conduction may be influenced by their different performance with locally generated noise. The main interference

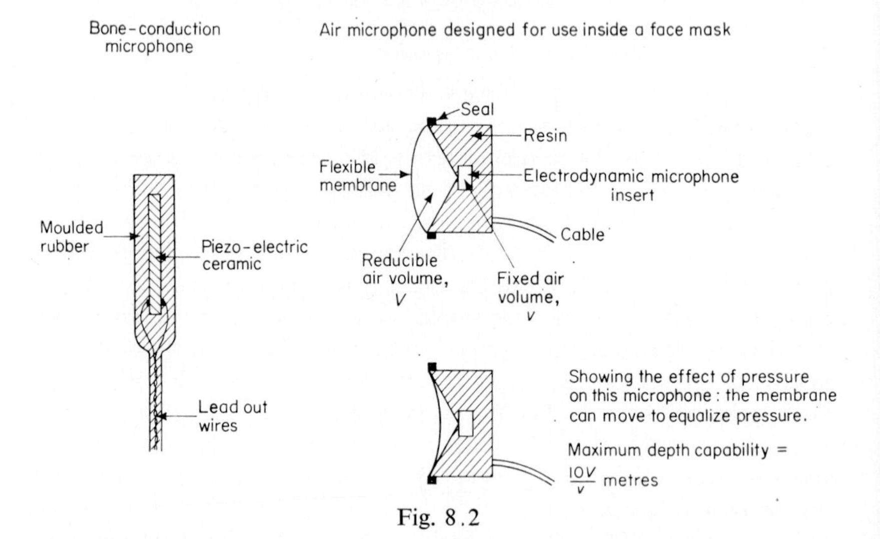

Fig. 8.2

experienced when using an air microphone is the sound generated by the breathing set, particularly the inhalation noise. On the other hand it is the noise caused by the exhalation bubbles that is the main limitation of the bone-conduction microphone.

8.3.1 Transmission through water

Referring again to Fig. 8.1 it can be seen that the stage has now been reached where an electrical representation exists of the words spoken by the transmitting diver. Some methods that can be used to transmit a speech signal through water are listed in Table 8.2. The simplest, and often most efficient of these is to use a pair of wires between the transmitter and the receiver. If it is convenient to combine the telephone line with an air- or life-

Table 8.2 *Telephony transmission systems*

Method	Propagation	Radiator	Receiver	Main limitation	Typical range
Two wires	Electric current	Telephone	Telephone	The inconvenience of the wires	No limitation
Direct audio	Audio frequency sound	Piston sound (loudspeaker)	The ear	Noise generated by the receiving diver and his breathing set	0–30 m
Ultrasonic	Modulated high frequency sound 8–200 kHz	Resonant piezo-electric transducer	Resonant piezo-electric transducer	Reverberation and multi-path distortion	0·1–5 km
Electric field	(a) Audio frequency electric current field (b) Modulated high frequency field	Two spaced electrodes	Two spaced electrodes somewhat aligned to the transmitting electrodes	Interference due to any electrical installation with earth currents	A few metres with early models. Information rather obscure on later ones
Optical	Visible light (green window)	Laser or semi-conducting diode	Photodiode or photocell	Scattering and absorption of transmitted beam. Transmitters and receivers are generally highly directional	0–100 m (clear water) 0–5 m (coastal water)
Magnetic field	Magnetic induction	Large coil	Coil somewhat aligned to transmitting coil		
Very low frequency radio	EM waves	Aerial (impractically large for diver)	Aerial		

line, then this method is the obvious choice. However when the diver is not attached to any form of line a wire-less system is required.

The simplest method of wire-less communication is to amplify the voice of the diver and to transmit it through the water in such a way that the unaided human ear can be used for reception. This has been described by the term 'Direct Audio', and can be considered as the equivalent of an ordinary public-address system. The advantages of a direct audio transmission are that it is simple, and that a two-way conversation is possible without the need for a send/receive switch. On the debit side, the range is limited and the performance is dependent on the protective clothing that the diver is wearing around the ears.

The most popular method of transmission is to modulate an ultrasonic carrier. Ordinary amplitude modulation (AM), frequency modulation (FM)[7], and single sideband AM systems are all in use, and there have been experiments with pulse modulation. The choice of modulation is governed by its ability to resist multiple-path distortion as this will be the predominant interference in water depths of interest to divers. Frequency modulation has a natural resistance to the arrival of spurious multipath signals that are of small amplitude compared with the carrier, high-amplitude signals will cause serious distortion. In contrast an AM system will demodulate all the incoming signals and present the listener with the reverberation. The ear is not unaccustomed to dealing with a reverberant signal and there will be further discussion on this later.

Although some of the first communication sets to be marketed used the electric conduction field set-up by two spaced electrodes on the back of the diver, and electric fields are generated by certain fish, this method is less popular today. The advantage of being free from acoustic limitations was offset by limited range and 'dead spots' when the transmitting and receiving 'aerials' were at right angles. The introduction of high-frequency carrier conduction field sets with automatic gain control at the receiver may well revive this technique.

8.4 Reception and noise

Diver-to-diver communication is inherently a short-range technique. Range limitations that are due to poor signal to ambient noise (sea-state, thermal, biological, etc.) ratio are seldom of interest; it is, in general, the acoustic noise locally generated at the receiving end that limits the performance of these diver communication systems. The major part of this often stems from the breathing equipment; however, this may not present as serious a problem as might be expected from acoustic measurements. Because the act of breathing may often completely mask incoming signals, it may be necessary to synchronize the breathing of the two divers; this is often helped by the valve noise received from the other party. With most forms of face mask the auditory canals are closed by the presence of water. This condition gives rise to an improvement in hearing by bone conduction of between 15 to 20 dB[8]. The effect of this is that body movement sounds and noise generated by movement of clothing and equipment on the body become significant underwater. A simple example of this is the effect of scratching the back of the head; in air this is unlikely to interfere with normal hearing, whereas underwater, this may well mask voice communication.

Table 8.3 *Threshold of hearing underwater*

Comparison of published figures at 2 kHz. The divers are not wearing a protective rubber helmet.

Authors	Reference no.	Value of threshold ref. 0·0002 dyne cm^{-2}
Ide, J. M. (1944) (obtained from Ref. 15)	—	73 dB
Hamilton, P. M. (1957)	14	53 dB
Wainwright, W. N. (1958)	15	82 dB
Montague, W. E., and Strickland, J. F. (1961)	10	70–80 dB
Brandt, J. F., and Hollien, H. (1967)	16	60–70 dB
Zwislocki, J. (1957)	17	Threshold of bone conduction hearing in air 46 dB

There is a further source of noise that is significant in air-filled structures such as sea-bed shelters and underwater laboratories. By nature of their design these structures are often highly reverberant; this is particularly so in the case of some of the smaller flexible ones. Here the 'Q' of the gas-filled shelter may approach that of a bell on land. Excess gas is often continuously vented from such structures and the release of bubbles appears to be an excellent way of exciting their resonances. The noise level inside can be sufficient to mask telephone conversation and to cause considerable annoyance to the occupants.

8.4.1 Hearing

The diver must equalize the pressure across the ear 'drum'. This is normally accomplished by allowing the outer ear to be open to the water or helmet cavity, and equalizing the pressure in the middle ear by opening the eustachian tubes periodically (swallowing or blowing into the nose are the usual methods). Failure to equalize properly can cause loss in hearing sensitivity, and complete neglect can lead to rupture of the drum[9].

Normally when man swims under water there is a bubble of air trapped in the auditory canal; the part played by this is uncertain, as is the mechanism by which sound reaches the inner ear. It is possible that almost all hearing underwater is by bone conduction, and the similarity between the underwater and bone-conduction hearing thresholds have been used as evidence to support this[10]. Alternatively the sound may be transmitted along the auditory canal in much the same way as in air; the ability to hear directional information is the evidence in this case. Table 8.3 compares some of the published values for underwater thresholds at 2 kHz. The author is of the opinion that it is not meaningful to consider underwater hearing as either tympanic or bone conduction. These terms apply to hearing in air where the difference in acoustic properties between human flesh-and-bone, and gas, is such that one can consider a sound wave as travelling in one or the other. Under water it may well be more reasonable to consider the two hearing organs as existing in an infinite fluid. This ignores the water–skin boundary completely. The middle ear, sinuses and other air cavities are impedance discontinuities in the neighbourhood of the receptors and must be taken into account. Directional hearing and changes in angular sensitivity can be explained while retaining some similarity with bone conduction hearing.

It is possible for an underwater swimmer to tell the direction from which a sound originates[11]. This suggests that there may be considerable advantages to be gained in underwater communication by using a binaural receiver, as it is well known that binaural hearing in air helps in the understanding of a speech signal in the presence of noise or reverberation. This ability is well demonstrated by the 'cocktail-party effect'[12]; a person 'listening' to several conversations simultaneously has little difficulty in isolating them and understanding whichever one he pleases. This ability, however, is lost when listening to a monophonic recording of the same sound.

Although the use of binaural receivers has been proposed before[13], the only available equipment that allows binaural reception are the 'direct audio' communicators. These, of course, use (both) the unaided ear(s) for reception. It has been reported[3] that under some conditions communicaion is possible with 'direct audio' equipment where carrier systems failed due to reverberation.

8.5 Conclusions

In the experience of the author, most equipment designed for diver communication fails through mechanical design, either leakage (poor seal design), corrosion (wrong combinations of metals), or it encumbers the diver (buoyancy index, physical size or method of operation). Having eliminated equipment that cannot be reliably tested outside the laboratory, the second most common failing is in the acoustic design at the transmitting end: facemask, microphone and breathing set.

It is only after the mechanical and acoustic design has been considered need one look to the ability of the equipment to transmit the voice signal through the water; yet it is in this area where almost all the evaluation of devices is done. It is of little wonder that a communicator that performs well in the test tank all too often fails to be of practical worth.

Although it has been emphasized that the whole of the diver's equipment must be designed with communication in mind, it is, of course, well known that the human brain is very tolerant to a considerable amount of speech distortion. It is possible to make up for a weak link in the communication chain providing all the other links are capable of passing on the distortion produced in order to give the person at the receiving end the best chance of 'reading' the signal. For this reason, helium speech plus a pair of wires, a poor facemask plus a pair of wires, or a person speaking in air with a poor submarine telephone, may all be intelligible. Whereas combinations that use two weak links are seldom useful; their encumberance generally outweighs any communication advantages they may offer the diver.

References

1 O.N.R. Report A.C.R.-124, *Project Sealab Report*, p. 143, 'Swimmer Intercom' Office of Naval Research, Washington, D.C. (1967).

2 Stover, W. R., 'Technique for Correcting Helium Speech Distortion', *J. Acoust. Soc. Am.* **41**, 70 (1967).

3 Ray, B., 'Voice Communication Between Divers', *Underwater Association Report 1966–67*, p. 47, Iliffe, London.

4 Stephens, R. W. B., and A. E. Bates, *Acoustics and Vibrational Physics* **15**, 18, Arnold, London, 1966.

5 Needy, K. K., 'Divers Communication Improved', *Science* **153**, 321 (1966).

6 Webb, H. J., and J. R. Webb, 'An Underwater Audio Communicator', *Inst. Elec. Electron. Engrs Trans.* **AU-14**, 127 (1966).

7 Gazey, B. K., and J. C. Morris, 'An Underwater Acoustic Telephone for Free-swimming Divers', *Electron. Eng.*, June, 1964.

8 Zwislocki, J., 'Ear Protectors', *Handbook of Noise Control* (C. M. Harris, ed.), pp. 8–12, McGraw-Hill, New York, 1957.

9 Miles, S., *Underwater Medicine*, p. 85, Staples, London, 1966.

10 Montague, W. E., and J. F. Strickland, 'Sensitivity of the Water Immersed Ear to High- and Low-level Tones', *J. Acoust. Soc. Am.* **33**, 1376 (1961).

11 Ray, B., 'Communication Between Divers', *Oceanology International 69* (Proceedings of the Society for Underwater Technology Conference) (1969).

12 Cherry, C., *On Human Communication*, Ch. 7, 4.3, Science Editions, New York, 1961.

13 Bauer, B. B., A. L. DiMattia and A. J. Rosenheck, 'Transmission of Directional Perception', *Inst. Elec. Electron. Engrs Trans.* **AU-13**, 5 (1965).
14 Hamilton, P. M., 'Underwater Hearing Thresholds', *J. Acoust. Soc. Am.* **29**, 792 (1957).
15 Wainwright, W. N., 'On Comparison of Hearing Thresholds in Air and in Water', *J. Acoust. Soc. Am.* **30**, 1025 (1958).
16 Brandt, J. F., and H. Hollien, 'Underwater Hearing Thresholds in Man', *J. Acoust. Soc. Am.* **42**, 966 (1967).
17 Zwislocki, J., 'In Search of the Bone-conduction Threshold in a Free Sound Field', *J. Acoust. Soc. Am.* **29**, 795 (1957).

Author Index

Subject Index